PRAISE FOR *THE MATHS OF LIFE AND DEATH*

'An exciting new voice in the world of science communication'

Marcus du Sautoy

'This is an exquisitely interesting book. It's a deeply serious one too and, for those like me who have little maths, it's delightfully readable'

Ian McEwan

'An extremely thoughtful, articulate and accessible insight into mathematics in the real world'

Alex Bellos

'A lucid and enthralling account of why maths matters in everyone's life. A real eye-opener'

Professor Ian Stewart FRS, author of *Do Dice Play God*

'Kit Yates is a brilliant explainer and storyteller'

Steven Strogatz

'A fascinating glimpse of mathematical epidemiology. A dizzying, dazzling debut'

Nature Magazine

'At times witty, at times charming, at times sombre, but always personable'

Aperiodical

'Quotable morsels of wisdom in every chapter'

Mathematical Gazette

HOW TO EXPECT THE UNEXPECTED

HOW TO EXPECT THE

The Science of
Making Predictions
and the Art
of Knowing
When Not To

KIT YATES

UN
EXPECTED

QUERCUS

First published in Great Britain in 2023 by

QUERCUS

Quercus Editions Ltd
Carmelite House
50 Victoria Embankment
London EC4Y 0DZ

An Hachette UK company

A CIP catalogue record for this book is available
from the British Library

HB ISBN 978 1 52940 867 6
TPB ISBN 978 1 52940 868 3
Ebook ISBN 978 1 52940 870 6

10 9 8 7 6 5 4

Typeset by CC Book Production
Printed and bound in Great Britain by Clays Ltd, Elcograf S.p.A.

Papers used by Quercus are from well-managed forests and other responsible sources.

For Emmie and Will – the best way to predict the future
is to create it for yourselves.

CONTENTS

Introduction

EXPECTING THE UNEXPECTED

Ever since the dawn of human civilisation, we have been trying to make predictions about the world and what it has in store for us. For just as long, we have been getting it wrong. Apocalyptic prophecies are a dramatic and surprisingly common example of such predictions, despite the unerring failure of all such forecasts in the past.

The Aztecs believed that four worlds had already been destroyed by the gods Quetzalcoatl and Tezcatlipoca, and that a fifth (ours) would be ripped apart by a catastrophic earthquake should they stop making human sacrifices to the gods. Suffice to say that as the Aztec empire waned, and their sacrifices dwindled, the world carried on regardless. The Hebrew book of Daniel, written around 165 BCE, predicted a catastrophic comeuppance for the Jews' Greek oppressors 1290 days after the Greeks had desecrated a Jewish temple. When this didn't come to pass, the last line of the book of Daniel was changed to make it 1335 days – but a month and a half later, still nothing had happened. French bishop Hilary of Poitiers (whose given name ironically means cheerful) pessimistically prophesied the End of Days for 365 CE, but when, embarrassingly, this didn't happen, his student

Martin (later to become Saint Martin of Tours) pushed the date back to 400 CE – another failure. Martin's successor and biographer, Gregory of Tours, at least had the good sense to predict a doomsday date between 799 and 806 CE, which would fail to come to fruition long after his death.

More recently, evangelical preachers like Harold Camping have made a good living out of predicting the rapture. Camping first calculated the 'End Times' to fall on 6 September 1994, but when it failed to materialise, pushed it back to 29 September and then 2 October. Surprisingly, after these humiliations in the 1990s, Camping received millions of dollars in donations from people who bought into his revised prediction of 21 October 2011. Camping and a range of other scaremongers received the 2011 Ig Nobel prize in mathematics (a satirical prize awarded for research that 'cannot, or should not, be reproduced') for 'teaching the world to be careful when making mathematical assumptions and calculations'.

Basing their predictions on little or no scientific evidence, it's not really surprising that these religious oracles ended up falling into the holes they had dug for themselves. Over the years, however, there have been some laughable predictions made by people who really ought to have known better. In 1830, in the earlier days of the railway era, science populariser and Fellow of the Royal Society, Dionysius Lardner, predicted that 'Rail travel at high speed is not possible because passengers, unable to breathe, would die of asphyxia'. This improbable warning was laughable, even at the time. Other predictions, however, have seemed funny only in hindsight.

When advising Henry Ford's lawyer on his plan to invest in the burgeoning Ford Motor Company in 1903, the president of the Michigan Savings Bank admonished, 'The horse is here to stay but the automobile is only a novelty – a fad'. In 2007, Microsoft CEO, Steve Ballmer, claimed, 'There's no chance that the iPhone is going to get any significant market share. No chance.' Yet other predictions are tragic in their naïvety or their wilful blindness to the inevitable. In September 1938, Neville Chamberlain returned from a meeting with Adolf Hitler claiming, 'For the second time in our history, a British prime minister has returned from Germany bringing peace with honour'. The Second World War began less than a year later.

Predicting the future is fraught with danger. No one wants to be the doomsayer whose apocalyptic predictions never materialise, leaving them a laughing stock. In 1970, American scientist James P. Lodge Jr, based at the National Center for Atmospheric Research in Boulder, Colorado, put himself in that position by proclaiming that 'Air pollution may obliterate the sun and cause a new ice age in the first third of the next century'. In 1971, Lodge's assertions were backed up by Columbia University's S. Ichtiaque Rasool and Stanford's Stephen H. Schneider, who claimed in the prestigious journal *Science* that rises in atmospheric dust over the next fifty years 'suggest a decrease in global temperature by as much as 3.5°K.' 'Such a large decrease . . .' they went on to claim '. . . is believed to be sufficient to trigger an ice age'.[1] Suffice to say this prediction did not come true. In fact, as we are only too aware, we face quite the opposite problem to global cooling.

At the other end of the spectrum, no one wants to stand in the shoes of British weather forecaster Michael Fish, giving a nation the all-clear in the face of an imminent disaster. During one forecast in October 1987, Fish confidently assured an anxious British public, 'Earlier on today, apparently, a woman rang the BBC and said she heard there was a hurricane on the way. Well, if you're watching, don't worry – there isn't.' The storm that hit the UK that evening was the worst for hundreds of years. Gales reaching 115 miles an hour wreaked devastation across the south of England, causing £2 billion of damage and killing eighteen people.

Despite the dangers of predicting the future, predict it we must. At a personal level, we need to know what the weather will be like this afternoon, so we can decide whether or not to hang out the washing; we need to know how heavy the traffic will be, so we can set off in time to make that important meeting; and we need to estimate our expenses, so that we can budget appropriately. These are the mundane day-to-day predictions that help our lives to run more smoothly and can cause us difficulty when we get them wrong.

On a larger scale, for the good of our broader society, we need to be able to predict and intervene to avert economic downturns; to forecast and prevent terrorist attacks; and to understand the current and potential threat of climate change in order to take action. If we fail in these high-stakes predictions, livelihoods, lives and even the fate of our species could be on the line. If we neglect the lessons taught by past experiences and fail to make sufficiently considered predictions, then we are likely to run into unanticipated scenarios: the firearms buy-back programmes that led to a rise in gun ownership, the safety features on cars that caused more deaths than they prevented or the species introduced to control a pest which ultimately became a scourge themselves.[2]

Many ways to be wrong

As well as highlighting ways to make better predictions to help future-proof our lives, this book is about the many different ways in which predictions can be wrong and the lessons we can learn to put them right. I will synthesise results from my native areas of mathematics and weave them together with studies from biology, psychology, sociology and medicine, theories from economics and physics, and, most importantly, experiences from the real world, to help you learn to expect the unexpected.

Two of the most important confounding phenomena that we experience routinely in our everyday lives and which we struggle to properly comprehend are *probability* and *nonlinearity*. We are not innately equipped to peer through clouds of uncertainty or to see what is coming around a bend in the road. Consequently, I will argue throughout that mathematics should be at the heart of our attempts to predict, for one very simple reason: mathematics can provide us with the objective tools to bypass the foibles of our own biology – the limitations imposed by our own thought processes, the compulsions that ultimately make us human, but let us down when it comes to making inferences about the world around us. Some of these ingrained impulses result from too much experience of certain phenomena and others from too little. They are humanity's shortcuts: the preconceptions and cognitive biases, refined over millennia of evolution, that all too often lead us astray when we try to apply our brains' old rules to our societies' new environments.

For example, one of the things my kids like to do on a nice day is to play on the trampoline. As I work in the garden, they are always begging for me to join them to act as mediator or to take part in the new games they are constantly making up. Whatever form the game takes at its beginning, it almost invariably ends up in a protracted wrestling match. When we're all too tired to go on, the three of us usually end up falling on to our backs, panting and looking up at the sky. This is secretly my favourite part of the proceedings, not just because I get a rest, but because it often heralds the start of a new and calmer game. We will look up at the clouds passing overhead and start to call out what we see. 'Can you see that turtle flying by over there?' one of them will point out. 'What, you mean the mermaid smoking a cigar?' I'll say. 'No, can't you see it's a dragon in a top hat?' the other one replies.

Cloud spotting is an old and ubiquitous game that relies on an old and ubiquitous habit. Our species' time-tested and universal ability to pick up patterns in noisy environments is sometimes referred to as *patternicity*. For example, many different cultures have evolved the 'man-in-the-moon' tradition, supposing they can make out a face or even the whole body of a person trapped in the irregular shadows of the lunar surface. This universality is likely a result of the fact that picking out human faces and figures from the background has always been an important skill for our species. Being able to recognise faces, for example, and to rapidly read their emotions had the advantage, in the distant past, of allowing us to quickly distinguish a potentially threatening individual and to read their mental state, so that we might prepare for flight or a fight. Neurologically, we have become hard-wired to pick out faces; there is even an area of the visual cortex – the fusiform face area – responsible for recognising and remembering them.[3]

Nowadays, spotting Jesus in a piece of burnt toast makes for a diverting newspaper story, but sometimes, these long-honed pattern-recognition skills, enabling us to find order in the midst of disorder, can lead us to jump to the wrong conclusions. Gamblers may believe they've picked up a pattern in the numbers in the lottery or on the roulette wheel, when clearly no such pattern exists. Investors may convince themselves they have developed a system to beat the market, when all they've really done is spot a non-existent trend in the messy stock trajectories. Scientists may find a cluster of disease cases and conclude there is a specific environmental cause, when in fact, the cluster occurred by chance, as a result of the random distribution of such cases, and no such connection exists. These sorts of mistakes, which we will explore more deeply in Chapters 2, 3 and 4, are a direct consequence of our inability to reason in the face of randomness and uncertainty.

Certainly uncertain

When it comes to talking about uncertainty, it's important to be clear early on that prediction isn't just about fathoming the future. There are things in the present about which we are uncertain. Indeed, there are a multitude of phenomena in the past for which we don't have the full picture. When the Irish archbishop James Usher offered up the spuriously precise and spectacularly wrong date of 22 October 4004 BCE for the day of earth's creation, he was making an incorrect prediction about something which had already happened. Economists are only too aware of this problem. By the time they've gathered the data on the metrics that tell us we're about to enter a recession, we're usually already there. To get an accurate picture of what's happening now, economists gather data from the more distant past in order to make 'nowcasts'[4] about what has happened in the recent past and what is happening in the present, for which we don't yet have data. Using similar methods, health researchers feed social-media data into nowcasting models to detect flu epidemics,[5] which haven't yet been picked up by health officials.

So there are actually, roughly speaking, two types of prediction which deal with the two types of uncertainty we come across daily: *aleatoric* (from the Latin *aleator* meaning dice player) and *epistemic* (from the Ancient Greek *episteme* meaning knowledge or science). To illustrate the difference, imagine I have a fair die in my hand. I ask you what the probability is of it coming up as a six when I roll it? No doubt, you will quickly tell me it's one in six. Ignoring the potential for the die to be biased, one in six is correct and reflects your *aleatoric* uncertainty: uncertainty in the face of an event involving outcomes that can differ each time the experiment is performed. Now I ask you to turn around while I roll the die and cover it with my hand before you turn back. After rolling, when I ask you what is the probability that the die under my hand has come up as a six, how do you answer? Grudgingly, you will probably still tell me it's one in six. And you would be right again. This time, though, your answer reflects the *epistemic* uncertainty that occurs when we are asked to reason about a phenomenon already in existence or a situation which has previously played out, but of which we lack perfect knowledge.

Different fields of enquiry have their own slightly nuanced variations on the definitions of these two terms – epistemic and aleatoric – but for our purposes, these characterisations will suffice. For us, playing the lottery is a game of aleatoric uncertainty, as the random event of the draw has yet to play out. Buying a scratch card, however, is very much an exercise in epistemic uncertainty – gambling on the predetermined, but as yet unknown pictures that lie beneath the scratch-off panel.

For as long as we've been dealing with the aleatoric uncertainty inherent in forecasting the future, we've been demanding answers to epistemic questions about the nature of reality. The ancient Egyptians believed the earth was a flat disc.[6] Many of the ancient Greeks agreed. So, in one form or another, did the ancient Hindus, Buddhists, Mesopotamians, Chinese and most of the other ancient civilisations that had even thought to ask the question.

It was well into the Middle Ages before the spherical view of the world became the predominant theory. When Columbus set sail for Asia in 1492 (a journey which would eventually land him in America – so much for that prediction), some people still believed he might sail right off the edge of the earth. It wasn't until the Portuguese explorer Magellan completed his first circumnavigation of the earth thirty years later that the issue was definitively put to bed. Suggesting falsifiable hypotheses about the nature of our existence, like the earliest proposals by Pythagoras that the world was not flat, is the basis of the scientific method. It is the only reason we know anything about anything. Scientific theories are nothing more than epistemic predictions about the nature of reality that haven't been proved wrong.

The two types of uncertainty are not mutually exclusive. Many events in which randomness plays a key role will have elements of both, as we will discover in Chapters 2, 3 and 4. For example, in 2011, when Barack Obama gave the go-ahead for a team of Navy SEALs to attack the compound in Abbottabad in which he believed Osama Bin Laden was hiding, he was not sure of the mission's success. In an interview after the event, Obama candidly acknowledged the two separate sources of uncertainty with which he was faced. With the spectre of previous botched military interventions (including the 'Black Hawk Down' and the Iranian hostage-rescue incidents) weighing heavily on his mind, Obama said of the aleatoric uncertainty involved: 'There are a lot of things that could go wrong . . . There are huge risks that these guys are taking . . . These are tough complicated operations.' Separately, Obama acknowledged that the evidence he had been presented with to demonstrate that it truly was Bin Laden himself hiding out in the Abbottabad complex was far from conclusive. 'We could not say definitively that Bin Laden was there. Had he not been there, then there would have been significant consequences.' He characterised his perception of the likelihood of Bin Laden's residence within the complex – his epistemic uncertainty about an unknown fact – as 'a 55/45 situation'.

Linear solutions to nonlinear problems

So our brains are capable of making overgeneralisations and oversimplifications when asked to deal with probabilities, but it's worth remembering that we take other potentially injurious shortcuts, even when reasoning about scenarios in which uncertainty is seemingly absent. One of the most important cognitive economisations, whose ubiquity I expose in Chapter 6, is *linearity bias* – the propensity to believe that things will stay constant or continue to change at a consistent rate. Putting a fixed amount of our pay packets under the mattress each month means our savings increase linearly. If you get paid an hourly rate, then your pay packet increases linearly with the hours you work. If you work a little more one week, a fixed increase in the time you work should correspond to a fixed increase in your pre-tax pay. For linear processes, a fixed change in the input should correspond to a fixed change in the corresponding output. But many of the processes in our world are not linear. Nonlinearity, which we will discover in the later chapters of this book, is the second confounding factor (alongside probability) that foils our naïve attempts at prediction.

Our relative underacquaintance with nonlinear processes means their influence can surprise us. In Chapter 6, we'll meet the reciprocal relationship between fuel consumption and fuel efficiency that can fool us into making bad environmental decisions. While in Chapter 7, we'll see how the exponential growth of infected people at the start of an epidemic can catch us off guard, seemingly moving from manageably and reassuringly slow to alarmingly and unexpectedly fast. Even the quadratic relationship between the diameter of a pizza and its area can leave us out of pocket if we're not careful.

As a case in point, sometimes when I'm driving along the motorway on the long journey home from work, the idea of increasing the speed I'm driving at to get home sooner and to spend more time with my family is appealing. But every time I'm tempted, I remember that a fixed increase in the speed I'm travelling at doesn't give me a fixed increase in the time I'll save. The relationship isn't linear. Going over the speed limit on the motorway really isn't worth the risk of being caught. Changing speed from 50 to 70 mph saves you four and a half minutes over a 10-mile journey. However, stepping up another 20 mph, to 90 mph, saves you less than two minutes more over the same journey. With this simple nonlinear relationship, there are diminishing time returns the faster you travel. By considering everyday examples like these, I will highlight the simple cognitive shortfalls we are subject to and arm you with the ability to recognise them in and for yourself.

Sometimes, our overfamiliarity with some experiences can cross-react with our lack of exposure to others – particularly those for which complex dynamical behaviour and uncertainty are inherent – to leave us feeling powerless when we come face to face with an unusual scenario. The phenomenon of *normalcy bias* results from just such a combination of our familiarity with linear relationships and our unfamiliarity with extreme events. We assume that things will continue linearly – just the way they are now. It causes us to mini-mise, question or disregard warnings of imminent threats, because they are so far beyond the realm of our experience as to make them unbelievable.

The loss of life on the *Titanic* is often proffered as a prime example of people exhibiting normalcy bias. In the hours after the ship hit the fated iceberg, not all the *Titanic's* passengers treated the collision with the due reverence it deserved. Many had been led to believe that the ship was unsinkable. Even after receiving reports of the sinking, Philip Franklin, the vice president of the White Star Line (the ship-ping company that operated the *Titanic*), told the relatives and friends of passengers, as well as the assembled press in New York, 'There is no danger that Titanic will sink. The boat is unsinkable, and nothing but inconvenience will be suffered by the passengers.'

Many of the passengers tragically believed too strongly in the 'unsinkable' rhetoric, preferring the safety and relative comfort of the ship they had sailed on for days to the prospect of launching out into the unknown dark and freezing waters of the Atlantic in the middle of the night. Many of the early lifeboats that launched were not filled to capacity – not because they launched too quickly, but because people were hesitant to come forwards when called. Even those who did occupy the lifeboats when they were lowered down from the deck of the ship recall being sceptical as to whether this 'precaution' was really worthwhile. People just couldn't believe that their deeply held expectations of the next few days – sailing safely across the Atlantic to New York on a ship which Captain John Edwards claimed 'Even God himself couldn't sink' – could be jeopardised in this way. Many passengers took too long to give up on that comfortable version of their future, even in the face of the stark warnings they were given.

As it transpired, there were not enough lifeboats for all those on board the *Titanic* in any case: a White Star Line decision borne out of a misplaced confidence that the ship would not sink and an aesthetic desire to maximise the deck space available for the passengers to enjoy. This complacency, combined with the underfilling of the early lifeboats meant, tragically, that many lives were lost that night to the cold, cruel Atlantic Ocean – lives which might otherwise have been saved if it weren't for normalcy bias. We will discover more of the pernicious impacts of normalcy bias in Chapter 9.

The nonlinear phenomena described thus far – reciprocal, exponential and quadratic relationships – are, as we will see in later chapters, among the easier concepts to get our heads around. Yet we persistently make mistakes with them. How then should we expect to make predictions about the behaviour of complex systems that are riddled with feedback loops, discontinuities, oscillations and other more complicated nonlinear behaviour, and which depend on many interdependent variables? Situations can run out of control beyond our expected horizons extremely quickly in these scenarios.

Maths offers the potential to act as a guide through this nonlinear world. Having the cold, hard logic of mathematics at our sides can help us to reason past the shortcuts our brains would intuitively like us to take. But even mathematics can only take us so far in the face of an inherently complex world. Even in systems for which we think we have eliminated uncertainty, there can still be inherent problems that mean we can't always say what will happen with perfect accuracy or arbitrarily far off into the future. Despite the undoubted successes of mathematical clairvoyance – predicting everything from the location of missing planets[7] to the existence of radio waves[8] – we often struggle to understand and predict seemingly simple phenomena: the pitter-patter of a dripping tap[9] or the fluctuations of animal populations.[10] If you've ever played Poohsticks, you'll know that roughly the same-sized sticks dropped in roughly the same place at roughly the same time can take extremely different paths, even over the short distance from one side of the bridge to the other. This is a caricature of *chaos*.

As I will highlight in the final chapter, chaos can stymie our attempts to make vital forecasts about what, in theory, should be predictable systems: the populations of endangered animal species, the trajectories of epidemic spread, the behaviour of crowds and, of course, the weather. Unpredictable behaviour can emerge from well-characterised systems, even in the absence of external sources of randomness.

Even the much-vaunted power of maths has its boundaries. There are fundamental restrictions which hamper our ability to foretell. Although mathematics gives us an unprecedented tool with which to project forwards, uncertainty and nonlinearity place definitive horizons on how far we can ever hope to see into the future.

Expecting the unexpected

As well as suggesting ways in which we can try to predict the future, perhaps more fundamentally, this book is about identifying and understanding the barriers we come up against when trying to do so. We can hope to learn something by reviewing the many and varied ways – sometimes amusing, sometimes tragic, but always relevant – in which our simple prognostications can and do fail: our 'gut feelings' – divinations based on supernatural or instinctual reasoning, which, purely by chance, can occasionally be correct (even a stopped clock is right twice a day), but miss the mark most of the time because they have no scientific basis; the 'everyday extraordinary' events that, at an individual level, seem so rare as to be almost impossible, but at a population scale become almost inevitable; the events which are 'inherent uncertainties', for which we can say much about their expected frequency in the abstract, but little concrete about the occurrence of any individual event; the 'common tragedies' – in which the short-term rational behaviour in the best interests of individuals can cause everyone in the group to lose out in the long term.

We must watch out for the 'curveballs' that seems to travel in a straight line, but veer away from their predicted trajectory at the crucial moment; the 'snowball' positive-feedback loops that, starting innocuously, can roll out of control, gathering mass, eventually becoming an avalanche; the 'boomerang' negative feedbacks – those predictions that can change the phenomenon about which they forecast, leading to a different result; and, finally, the 'fundamental limits' imposed on us by the very nature of the world in which we live, placing restrictions on how far into the future we can expect to forecast and the accuracy with which we can ever hope to do so.

Throughout the book I will provide insights and tips that will demonstrate how to avoid being taken in by unfounded predictions and allow us to figure out who to believe. I'll pick apart the folklore and rules of thumb we've been using to make predictions for centuries, explaining the science behind those 'red skies at night' and debunking myths like the 'lying-down cows'. I'll give you some of the tools to make your own predictions and help you learn when not to trust your basic intuition. We'll delve deep into the fabric of our reality, as we travel the path of reason that leads through the cloud of probability, and we'll shine a light on the situations in which something more than a verbal, linear argument is needed.

My fundamental task is to alert you to the many and varied ways in which predictions can go wrong: the ways in which your intuition can be fooled or your better judgment clouded by a seemingly convincing argument. More than just illustrating other people's mistakes, though, I will try to empower you to make future-facing decisions of your own by taking the simple tips and tools I provide, and using them in real-world scenarios.

There is no silver bullet for making accurate predictions for any and all scenarios – no telescope which allows us to see unencumbered into the far reaches of the future. Sometimes things happen which are genuinely impossible to predict. On other occasions, our actions today have far-reaching and unintended consequences for tomorrow. No mathematical formula or stack of data, no matter how well processed, will be able to sound a warning alarm with perfect accuracy.

However, there are plenty of scenarios in which we can make credible predictions about the future, but in which we fail to do so because we are either unaware of the instruments of forecasting or perceive that we lack the authority to wield them. This is what this book is all about – learning lessons from unsuccessful prognostications in the past and recasting these mistakes as an arsenal with which to make more reliable predictions about the future. By the end of the final chapter, you will be able to see more clearly into the mist of the seemingly uncertain events that lie in wait, as you begin to expect the unexpected.

1

GUT FEELINGS

It is unseasonably warm for October when I step off the dark, busy street in central London and into the small, brightly lit shop. Spiritualism specialises in a wide range of miraculous charms, including healing crystals, Ayurvedic tinctures and supernatural stones – enough to fill a quarry. You might be wondering what on earth I, a scientist and sceptic, am doing here? Safe to say, I'm not here to buy an amulet or a dream catcher. The magic rocks and other mystic debris that litter the shop seem unduly expensive, but where Spiritualism really makes its money (and what I'm interested in) is through delivering psychic readings – deciphering what the future holds in store for people or putting them in contact with 'the other side'. And when better to make contact, I reason, than the week before Halloween when, I am reliably informed, 'the veil between our world and the world of the spirit grows thin'. No doubt, they would have been able to forecast me walking in off the street for a reading, but just in case, I rang ahead the week before and made myself an appointment with Paula, one of the resident clairvoyants.

As I wait for Paula to materialise from her spiritual sanctuary (the basement), I shuffle nervously among the cramped shelves, stopping in front of each display in turn, half-reading the almost comically specific inscriptions. 'Bloodstone – a stone to overcome influences such as electromagnetic stress'. 'Bronzite – known to protect against curses'. 'Amethyst – guards against psychic attack.' If things go wrong downstairs, I may need this one.

Trying to foretell the future helps us to feel in control of its inherent uncertainties, to manage our aspirations and to make important decisions. To predict, even in the absence of evidence, is a natural human desire – a gut instinct. We've been using a variety of bizarre and unscientific methods to do this for millennia – none of them seemingly any more reliable than the others. Typically, our ancestors viewed their various methods of fortune-telling as a way of interpreting the will of their god or gods. It's no coincidence that 'divine', the verb (meaning to gain supernatural insight) and 'divine', the adjective (meaning associated with God or godlike) are near homonyms in many languages.

From as early as the tenth century BCE, the ancient Chinese used a divination manual, the *I Ching* or *Book of Changes*, to help them ascertain 'divine truth'. The practitioner would repeatedly cast yarrow stalks (or nowadays, typically coins) to generate a random series of six ones or zeros, which could then be converted into a pattern of broken (yin) or full (yang) lines known as a hexagram. The two equally likely choices for each of the six lines meant there were 2^6 – or sixty-four – equally likely hexagrams forming a binary code, as illustrated in Figure 1-1. Each hexagram represented a corresponding section in the text which could be interpreted by a skilled reader to make predictions or suggest future actions.

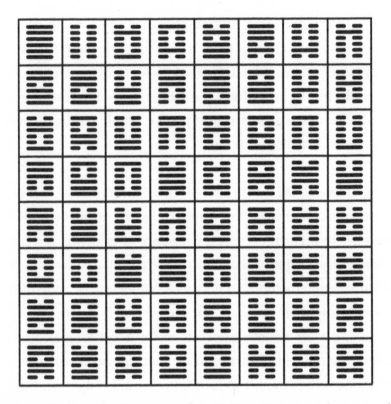

Figure 1-1: The sixty-four hexagrams of the I Ching. *From top to bottom, each of the six positions can be filled with either a full or a broken line. These two options in each of the six positions gives sixty-four (2⁶) equally likely possibilities.*

The use of objects to generate a random number or pattern, which is then interpreted by a well-versed 'seer', was a common theme among many early forms of divination. The practices are grouped under the umbrella term *cleromancy* – a Latinised combination of the Greek *kleros*, meaning lots (as in casting lots) and the suffix *mancy*, meaning divination. Cleromantic methods are among the oldest forms of divination and have arisen independently in many different cultures. In a similar vein to the method of consulting the *I Ching*, the Yoruba of West Africa practise Ifá divination. The Babalawos – priests of the Ifá oracle – cast kola nuts to make a series of eight broken or unbroken lines in a tray, which is traditionally filled with specially sanctified termite dust. The binary system of eight lines forms a code which indicates one of 2^8, or equivalently 256, tonal poems which can provide guidance about the future.

The casting of lots, be it rolling dice, flipping coins or drawing straws, is also part of the Judeo–Christian tradition. Perhaps most famously, after disobeying God's instructions and running away, Jonah (himself a prophet, whose self-defeating prophecies we will hear about in Chapter 8) finds himself on board a ship as a storm gathers in the ocean around it. Keen to find out whose god is responsible for the gale, the sailors on board cast lots and, so the story goes, 'the lot fell on Jonah'. This act of divine providence leads to him being thrown overboard and then swallowed by a big fish (or a whale, depending on who's telling the story).

Another way of introducing the randomness required to produce the 'unknowable-providence-of-the-divine' factor is the generation of unpredictable patterns. *Tasseography* (interpreting tea leaves) is a classic method of fortune-telling in which an unfiltered cup of tea is drunk down to the dregs. The tea leaves which settle on the sides and bottom of the cup form a pattern that, with the help of a vivid imagination, can be interpreted by the reader. Generic symbols such as arrows, moons and wheels have a variety of ambiguous meanings (change, news, success, etc.) in the tasseographer's interpretation manual, allowing the reader to tailor their predictions to what they perceive the drinker might want to hear. Older versions of a similar practice involved interpreting splatters and puddles produced by molten wax (*carromancy/ceromancy*) or lead (*molybdomancy*).

An altogether more gruesome way of finding an unpredictable pattern to interpret was popular among the ancient people of Greece, Italy and Mesopotamia, going back to at least the third millennium BCE. *Haruspicy* or *extispicy* involves examining and interpreting the entrails, especially the livers, of sacrificed animals – generating a literal gut feeling. Perhaps the most infamous example of advice based on haruspicy was given by the soothsayer Spurrina to Roman emperor Julius Caesar in 44 BCE. After finding that a bull that Caesar had sacrificed allegedly had no heart, Spurrina gave the emperor the deliberately vague warning that 'his life would be in danger for the next thirty days'. Had the prediction not come true, Spurrina could have pointed to the improved care that Caesar took over his safety to absolve himself of the bad prediction. As it happened, on the thirtieth day after the prediction was made, on the Ides (fifteenth) of March, Caesar was murdered by a group of his own senators. The high-profile nature of the success explains the commemoration of this solitary prediction in a Shakespeare play, while all Spurrina's other, perhaps less successful, predictions have been mysteriously lost to history. In Chapter 3, we will revisit in more detail this phenomenon of *reporting bias* – by which only successful predictions are remarkable enough to be immortalised and survive the test of time, while incorrect efforts fade into obscurity, giving an overinflated impression of a forecaster's accuracy.

Another form of cleromancy, popular among many ancient human traditions as a method of discerning the will of the gods, was *astragalomancy* or dice divination. The original dice used were not the regular, number-scored cubes we use in games of chance today, but unadulterated animal bones – specifically, the cuboid-like ankle bones, or astragali, of sheep, pigs, goats and deer. The precise rituals and games in which the dice were employed varied by tradition. As particular rolls became imbued with meaning, the faces of the dice eventually came to be marked with representative symbols. When used in divination, the associated marks thrown up by the cast dice could be interpreted by the diviner to answer their questions. These sacred games were the precursors of modern games of chance and, as people began to place wagers on the outcomes of the games, what we now consider gambling developed and merged with these spiritual practices.

The uneven outcomes of the astragali dice led to their being whittled into cubic shapes to form the first objects that we would recognise as the modern-day dice used in board games and on craps tables throughout the world. The study of the outcomes of games of chance involving dice led to the foundations of modern probability theory, which, as we will see in later chapters, is fundamental to modern methods of predicting the future.

While the uses of dice for divination predate their use as random number generators for games and gambling, the opposite progression holds true of playing cards. Playing cards probably had their origins in the ninth-century Tang dynasty in China. It wasn't until they spread westwards to Europe in the fourteenth century that *cartomancy* – fortune-telling using a deck of cards – began to gain popularity. Although now one of the more widely employed tools for divination, tarot cards only gained their occult connotations and thus became popular for divination during the eighteenth century. The traditional Italian suit of swords (English clubs) was rebranded as wands to provide a mystical air. Coins (English diamonds) were recast as magic-evoking pentacles. An extra twenty-two character cards, including 'the Magician' and 'the Emperor', were introduced, presumably to make remembering the supposed meanings of the cards a little bit easier. Tarot cards are shuffled to randomise the deck and the reader typically allows the sitter to choose a certain number of them, which can then be interpreted by the cartomancer to give a personalised message.

The illusion of randomness

The single theme that runs across so many of these early forms of divination, from *acultomancy* (interpreting the unpredictable patterns needles make when dropped in flour) to *zoomancy* (interpreting seemingly erratic animal behaviour) is that of randomness. Cards were shuffled and picked at random, dice were cast to generate random outcomes or coins flipped to dictate random pieces of scripture. But why is it that mathematical randomisation or natural randomness played such an important role in fortune-telling – and still does, even to this day?

Here's a game for you to play. It comes from the mathematician's stable of mind-reading tricks. This modern-day prediction starts out like one of those commonplace 'think-of-a-number' gimmicks, a favourite trope of mathemagicians. You're going to need to keep a running total and to complete the task as quickly as possible, so if you need to get the calculator on your phone out to do it, that's fine. Ready? Then here we go.

Think of a number between one and ten. Triple this number. Add twelve to this. Divide the total by three. Finally subtract the number you first thought of. Now you'll have a final number in your head. Remember this number. We're going to convert it into a letter according to its place in the alphabet using a simple numerical code as follows:

A = 1	E = 5	I = 9	M = 13	Q = 17	U = 21	Y = 25
B = 2	F = 6	J = 10	N = 14	R = 18	V = 22	Z = 26
C = 3	G = 7	K = 11	O = 15	S = 19	W = 23	
D = 4	H = 8	L = 12	P = 16	T = 20	X = 24	

Take the letter that corresponds to the number you ended up with from the first part of the trick and think of a country – any country you like – beginning with that letter. Now take the second letter of the country you thought of and think of an animal that begins with this letter. If everything has gone right, then I'll predict that the colour of the animal you thought of is grey. I'd even go so far as to predict that it's a grey elephant from Denmark!

Was I right? If not, then either you're one of the small minority of people who think outside this particular box, or you got the maths wrong. If you did think of the Danish elephant, then you might be wondering how I divined such a specific answer from your definitively random and uncontrollable input. And here's the rub. Of course, I wasn't able to genuinely manipulate your mind to make you choose a specific number to start with. That really was up to you. But it turns out I could manipulate this random input to turn it into anything I wanted to. The maths part is pretty routine. If you weren't so busy trying to quickly work out the sums, you would have seen that, by asking you to treble your number, add twelve and then divide by three again, all I was really getting you to do, in a roundabout way, was to add four to your original number. Once I ask you to then subtract the number you first thought of from the running total, you should be left with just the four that I circuitously asked you to add. No matter what number they start with, everyone should end up with four.

Once I've got you to the number four and hence the letter D, the rest of the trick relies on exploiting common biases. Most English-speaking people, when asked to think of a country beginning with D, will go with Denmark, especially under time pressure. Even with time to think, you might struggle to think of another country. If you came up with Djibouti or the Democratic Republic of Congo, then fair play to you. From the E in Denmark the next common bias will suggest elephant to most people. Again, eels and eagles are possibilities, but much less common.

Exploiting these common biases distances the trick from the maths and draws participants' suspicions away from the possibility that they have been duped by some numerical sleight of hand. Even though most of us don't believe in it, this makes the alternative possibility of mind-reading seem more plausible. The key to the wow factor in the trick lies in the illusion of choice presented by the ten possibilities for the initial number and the full twenty-six-letter alphabet code. At the end of the mathematics, however, although the participant is not aware, the agency of their choice has vanished like the apparition it was always planned to be. Once the randomness is gone, I can then exploit your cognitive shortcuts to give you a prediction which makes it appear that I have genuinely read your mind.

Randomness also lies at the heart of many ancient and modern clairvoyant practices, its disorienting effects exploited or neutralised using a variety of techniques. How better to demonstrate the capricious will of an all-knowing deity than by using a divination mechanism with an apparently unknowable outcome? Control of the prediction is seemingly passed out of the hands of the seer in a classic 'nothing-up-my-sleeves' magician's trick. The haruspex or the cleromancer, the Babalawo or the tarot reader, seem to cede all control to whatever randomising 'force' is guiding the objects of divination.

In reality, the randomness in the conjuring trick is harnessed by the showmanship that typically accompanies the predictive act. The sleight of hand comes in the interpretation of the signs, when the audience thinks the trick is over – after the die is cast. The randomness serves to allow the interpreter a blank page on which to impose the narrative that best convinces the prediction-seeker that the diviner's powers are genuine. It provides the distraction that allows the seeker to be fooled into thinking the message has been delivered and that the trick is over, when really the magic of painting a story on the randomised canvas is just about to begin.

There is some significant skill involved in this narrative-composition step – exploiting our cognitive biases to selectively highlight or downplay certain of the unpredictable signs that are thrown up in the randomisation process. Without this story-weaving expertise, soothsayers, mystics and mediums would surely not have persisted for so long and been afforded such important positions in many ancient societies.

Something for everybody

Evidence of the popularity of soothsayers dates back many thousands of years to the ancient peoples of Egypt, China, Chaldea and Assyria. However, with the ushering in of the Age of Enlightenment in eighteenth-century Europe, the fortune-teller's popularity waned and many of their ceremonies fell by the wayside, victims of the heightened suspicion with which these non-scientific practices were increasingly being viewed. This scepticism spread around the world, as the Europeans extended their colonial reach.

Today, many of us deride these shamans and their prediction practices as unreliable nonsense. But the amorphous desire to believe in extrasensory ability – some hazy awareness or reception of information perceived through means other than the usual senses – still finds form in the willing vessels of many modern-day 'believers'. A 2005 survey by the US polling company Gallup found that over a quarter of all Americans believed in clairvoyance, while over three-quarters believed in at least one of ten paranormal phenomena, ranging from telepathy to astrology[11]. So why, even in the face of modern scientific consensus to the contrary, do many still believe in the power of horoscopes, premonitions and 'psychics'?

Attempting to answer this question is exactly what draws me to Spiritualism to have my fortune told by Paula. I am here to learn the tricks of the trade and to understand the everyday psychological spells that are cast on the willing victims of psychic charlatans. If Paula ever had me under one of these spells, however, it is almost immediately broken as she leads me downstairs to her consultation room. Instead of a dark, mood-lit parlour filled with comfy recliners, crystal balls and soft, jangly music playing in the background, I am squeezed into a room not much larger than your average toilet cubicle. The lights are fierce and stark, the walls are bare and on the table between two upright chairs sits what looks like a pack of battered and frayed playing cards. I remember now that Paula is a tarot reader and these must be the tools of her trade.

We sit down and Paula asks me, 'What would you like me to look at for you?' I make up something vaguely convincing about discovering things that are buried, perhaps subconsciously, in my past and that might be restricting my outlook on the future. Paula hands the deck of tarot cards over to me and asks me to shuffle. She spends a few seconds spreading the cards out in a long line and asks me to select five at random. This is when I make my first mistake.

Since I have already randomised the deck by shuffling it, I figure it doesn't matter where I choose the cards from. I pick the first five cards from the right-hand end of the splayed deck and slide them face down over the table. Paula raises her eyebrows. My selection does not look very random to her. I remind myself that, just like the 10,000 people in the UK each week who buy the lottery numbers 1, 2, 3, 4, 5, 6 – correctly reasoning (as we will come to conclude in Chapter 3) that this combination is just as likely to come up as any other six numbers, but being forced to share the jackpot 10,000 ways if it ever does – being a mathematical smartarse is not in my best interests here. I make a note to myself to pick more 'randomly' next time.

As Paula turns over the cards and begins to tell me about 'the threads' she has 'gathered from the gloom of the past', it soon becomes apparent that she is conducting what is known as a *cold reading* on me. She doesn't have any background information, so she is relying on extracting information from me to build her predictions. Looking at the cards she has turned over, she begins by throwing me some compliments, telling me I am very 'intuitive' and very 'empathic', that I 'read people well'. These general platitudes, known as *Barnum statements*[12] (named for American businessman, showman and renowned psychological manipulator Phineas Taylor Barnum), constitute a common opening gambit for psychics and are clearly a safe place for Paula to start to learn more about me. Barnum, whose shows were filled with often elaborate hoaxes, is said to have claimed of his circus 'we have something for everybody'. His sentiment nicely sums up the idea of a Barnum statement – a general personality characterisation which could apply to almost anyone. Consider, for example, how well the following assessment captures your personality:

> *You have a great need for other people to like and admire you. You have a tendency to be critical of yourself. You have a great deal of unused capacity which you have not turned to your advantage. While you have some personality weaknesses, you are generally able to compensate for them. Disciplined and self-controlled outside, you tend to be worrisome and insecure inside. At times, you have serious doubts as to whether you have made the right decision or done the right thing. You prefer a certain amount of change and variety, and become dissatisfied when hemmed in by restrictions and limitations. You pride yourself as an independent thinker and do not accept others' statements without satisfactory proof. You have found it unwise to be too frank in revealing yourself to others.*

It sounds pretty accurate, right? In fact, these are just a bunch of Barnum statements strung together and designed to elicit the *Forer effect*,[13] a prevalent psychological trait in which the recipient of a general and vague personality assessment interprets it as if it were extremely personal and unique. The effect is named after psychologist Bertram Forer who, after administering a personality test to each of his thirty-nine students, gave them what he said was an individualised personality description based on their results. When asked to rate the accuracy of the description on a scale from zero to five, the students gave an average score of 4.3, indicating that they believed the depictions Forer had come up with matched their personalities extremely well. Only later did Forer reveal that he had given each student exactly the same characterisation, comprising many of the above statements, which he had taken directly from an astrology book.

Barnum statements and the Forer effect have found a new home in online personality quizzes, which tend to ask you several seemingly unrelated questions and then reveal which Harry Potter character you most resemble or which Disney princess you take after. When I took the Harry Potter quiz on Buzzfeed, I was told I was Hogwarts' headmaster Albus Dumbledore: 'You're wise, quirky and very trusting. You're loved and respected by everyone, but sometimes you put too much pressure on yourself to make everything right' – a classic set of Barnum statements, which I was happy to accept. The comments beneath the final screen of the quiz – including 'Wow, this is accurate' and 'This describes me perfectly!' – show the power of the Forer effect.

Another of the statements Forer picked out for his students was the following:

At times you are extroverted, affable, sociable, while at other times you are introverted, wary, reserved.

As well as being a vague Barnum statement, this description is also an example of what is known as the *rainbow ruse*. By giving statements which comprise two or more opposing aspects of a given emotion or experience, at least one of which almost everyone will have encountered at different times in their lives, the rainbow ruse is a comprehensive catch-all. The statements are designed to cover the whole spectrum of an emotion or character trait from positive to negative, just as the rainbow separates white light into the full spectrum of colours from red to violet. Confirmation bias does the rest of the psychic's work for them, as our brains choose the aspect or aspects of the statements which best apply to us.

When trying to diagnose potential 'emotional blockages' for me, Paula gives a fairly crude illustration of the rainbow ruse, telling me, 'There are times when you're happy and you're up here,' holding her hand up high, 'and other times when you're sad and down there,' holding her hand correspondingly lower. Who hasn't felt *both* happy and sad during their lifetime? I think, but I murmur my assent nevertheless.

Keeping the customer satisfied

After hearing a few of these blanket statements, it dawns on me that Paula's goal throughout this reading is not necessarily to deliver revelatory information, but rather to get me to agree with as many of her statements as possible – to convince me of her abilities, so that I might come back again, or at the very least not ask for a refund. Keeping things vague and broad is one way to achieve this. Another way is flattery. Generally, people like to hear themselves described in a positive light – that they are skilful or kind or fun to be around – and I am no exception. So when Paula tells me, 'You've got lovely energy; very deep, very connected to your emotions', I find myself nodding in agreement, even though I don't believe in supernatural energy. Paula reads my reaction and goes further, refining her guesses: 'I like what I'm being shown here because, yes, you have the spiritual connections, but you've got a very nice energy, it's very warm, it's very caring and nurturing of others as well.'

Her flattering ploy relies on a subconscious bias known as the *Pollyanna principle*[14] – the tendency of people to accept and recall positive feedback more favourably than negative. The phenomenon is named after Eleanor H. Porter's 1913 children's novel *Pollyanna*, in which the eponymous protagonist searches for something to be happy about in every situation she finds herself in. Even after Pollyanna has been hit by a car and lost the use of her legs, she decides to be happy that she had the use of them in the first place.

Scientists at the National Institute for Physiological Sciences in Japan have even managed to figure out neurologically why compliments make us feel good.[15] Participants in their experiments were asked to fill in personality questionnaires and to introduce themselves in a short video. They were then strapped into a functional MRI scanner and given feedback on their answers. In subjects given complimentary feedback, an area of the brain known as the striatum was clearly activated on the scan. This is the same reward centre that lights up when experimental participants are given basic sustenance like food and drink or even gifts of money. That result suggests that paying someone a compliment could be considered tantamount to emotional bribery.

When phrased in a specific way, an appeal to vanity can also act as a subtle tool to ensure compliance from a sitter. Statements like 'As an intelligent person you can understand what I'm talking about here' almost demand agreement. Denying the understanding of the psychic's point might be perceived as a tacit admission of stupidity. Even the benign 'Does that make sense?' that Paula reaches for after almost every statement leaves scant room for disagreement. There is little I could have misunderstood in 'You're open to spirit and picking up on messages', even if I don't believe a word of it.

This last statement is an example of another tactic that Paula employs. By complimenting my open-mindedness and even suggesting I have supernatural abilities of my own – 'Oddly enough, although this is your first time, I do need to say there are very strong spiritual connections around you' – she bequeaths me *psychic credit*. If Paula can convince her sitters that they too are endowed with psychic ability, then it becomes less likely they will question the methods employed or the conclusions drawn as part of the reading. Paula delivers a confirmatory example to bolster my psychic credit bank: 'It's like thinking of someone and then they contact you.'

Of course, this has happened to me and probably to you, too. As we will see in the next chapter, these sorts of coincidences are surprisingly likely. I have definitely answered the phone with 'I was just thinking about you'. However, this usually happens when I've been thinking about organising to meet up with a friend and we both realised we needed to get in touch with each other to sort out the details; or if I haven't spoken to someone for a while and we've both felt the absence of each other's company. Whoever calls first, the other will experience a pleasant feeling associated with the mild coincidence. The more people you stay in touch with and the more calls you make or texts you send, the more likely it is to occur. Indeed, it's one of the most commonly cited coincidences, which is why it is such a good candidate for Paula to choose in order to suggest I have extrasensory perception.

Magical thinking

In the trade, the experience of seemingly meaningful coincidences with no apparent causal connection is referred to as *synchronicity*. Psychologist Carl Jung first introduced the concept[16] in the 1920s and used it to argue that the causal effect was, in fact, paranormal activity. This is an example of so-called *magical thinking* – when the causal relationship between two linked events is not immediately apparent, our brains can be quick to infer unjustified meaning, as we'll see in more detail in the next chapter. In 'believers', the mistaken attribution of significance to these chance events can lead to the development of superstitions.

Many sportspeople and fans will be familiar with magical thinking in the form of pre-match rituals. Former Chelsea captain John Terry gradually acquired pre-match superstitions as his career progressed, including playing the same CD in his car on the way to the ground and using the same urinal prior to kick-off. Frequently, after a victory, Terry would remember something he did differently in the lead-up to the match and ascribe a causal effect of the action to the positive result. By the time his career with Chelsea ended, he was completing up to fifty routines before a match – so many that he struggled to remember them all.

Terry recalls losing his 'lucky' shin pads after defeat in a Champions League game at Barcelona's Camp Nou in 2004 and demanding the Chelsea staff search all over the 100,000-seater stadium. 'Those shin pads had got me to where I was in the game – and I'd lost them,' he later recalled. He believed that wearing those specific items had some positive influence on the way he and the team performed and that without them his luck would run out. 'I've had those shin pads for so long and now this is it, all over,' he remembers thinking. The shin pads were never found, so Terry was forced to borrow a spare pair from his teammate Frank Lampard. The first game with the new shin pads resulted in a resounding victory, casting doubt on the effectiveness and necessity of Terry's superstitious routines. Nevertheless, the borrowed shin pads took on a new, mystical significance as a result of the win, and from then on became Terry's lucky pair.

In the mid-1980s, Koichi Ono, a behavioural psychologist at the Komazawa University in Tokyo, became fascinated by how these sorts of human superstitious behaviours were formed. He carefully designed an experiment[17] which would demonstrate humans ascribing effects to actions they had taken without any plausible evidence of a causal link – the very definition of superstitious behaviour. Individual student participants were left alone in a room which contained a table, on top of which were three levers, and a counter on the wall designed to record the points 'scored' by participants. The only objective the students were given was to score as many points as possible. To let them know they had successfully scored a point, a light would flash on and a buzzer would sound. The delivery of a point in close proximity to an action they had just taken caused many of the students to ascribe meaning to the action and to repeat it in order to score more points. Unbeknown to the students, though, their actions had no influence on when the points were delivered. Some students developed consistent superstitious behaviours, even though their actions did not always bring about the reward of a point. Others developed more flexible lever-pulling routines which changed and adapted in response to the delivery of points. One student adopted some extremely elaborate behaviours. After the delivery of one point while her right hand was on the casing of the lever, she decided to jump on to the desk touching the counter, the light or the wall with the same hand in an attempt to secure more points. After ten minutes, she jumped down from the desk, but just as she did so, another point was delivered, so she switched her behaviour to jumping. One point was delivered as she jumped and hit the ceiling with her slipper, so she continued this behaviour until, after about twenty-five minutes, she gave up, exhausted. Just as John Terry found with his new shin pads, a new superstition, when appropriately reinforced, can supplant previous ones.

The development of superstitious responses is by no means unique to adults. In 1987, Gregory Wagner and Edward Morris, both researchers at the University of Kansas, conducted an experiment with three- to six-year-old children.[18] Each child in turn was left alone in a room with a mechanical clown designed to dispense marbles at random times. Told that if they collected enough marbles, they could exchange them for a toy of their choosing, three-quarters of the children developed some sort of superstitious response designed to elicit the clown to dispense marbles. Some children pulled faces at him, while others touched his face or danced in front of him. One small girl even determined that the best way to procure a marble was to kiss the clown on the nose.

The term magical thinking comes from the cognitive dissonance that often arises when we experience the finely honed craft of a good conjurer. When she saws her lovely assistant in half, the magician forces our brains to hold two contradictory views at the same time:

1. The assistant has been severed in two and people who are thus cleaved don't stay alive for long.
2. The assistant's face is smiling, and his legs are wiggling, indicating that he is very much alive.

When the brain misses the trick, one of the simplest ways it finds to resolve this uncomfortable situation, as it continues to be bamboozled by the next illusion, is to simply appeal to magic. If we can't figure out how the trick is done, then maybe, we reason, the conjuror really does have special powers.

The Baader–Meinhof effect

Psychics capitalise upon our proclivity for magical thinking in exactly the same way as magicians. By exploiting coincidences, which they rebrand as synchronicity, psychics deceive us into thinking they know things that they could have no reasonable way of knowing, creating a cognitive dissonance which they hope their audience will resolve by accrediting to them extrasensory powers. Another class of coincidences that psychics use to convince customers of their precognisance and to keep them coming back is brought about by the *Baader–Meinhof effect*.

If you've not heard of it before, the chances are you will hear of it again soon. The effect describes those occasions when you encounter a piece of unfamiliar information – an unusual phrase, word or name – and soon after stumble upon it again, perhaps multiple times over. It seems that the Baader–Meinhof effect acquired its name on a discussion forum in 1995, after participants realised there was no universal name for the phenomenon. It's likely that after first learning of the far-left West German terrorist group, the coiner went on to hear about it again and again in the space of a short period of time and gave the phenomenon the memorable moniker in the hope that it would invoke the effect itself.

The more recent (but less redolent) name given to the effect is the *frequency illusion*:[19] when you learn of something novel and it then appears to crop up in all manner of places with increased frequency. The more unusual and memorable the word or phrase, the stronger the effect. You ask yourself how it is possible that you've not come across this term once in your whole life, yet here it is three times in a week. The coincidence seems so incredibly unlikely that it can send you off in search of potentially specious logic in order to explain it.

In truth, the word or phrase probably isn't really appearing more frequently since you first become cognisant of it, and the first time you remember hearing it probably isn't the first time you actually encountered it. For the frequency illusion to work, the word or phrase you perceive as new needs to be memorable enough to stick in your mind – to be unusual sounding or accompanied by an interesting context which makes it stand out. Given how many words or phrases we are exposed to each day, it's not surprising that we frequently encounter repeated information. When repetition happens with already familiar words, it is rarely worth commenting on, if we even notice it at all. This might be considered a form of selective attention – our brains tending to filter out this 'uninteresting' information. However, a phenomenon known as the *recency effect* – an instance of the more general family of *availability heuristics* (which we will meet again in Chapter 9) – keeps freshly acquired observations and information at the forefront of our minds. It means we are biased towards recognising information we recently assimilated. Combined with confirmation bias – in this case your belief that you really are seeing this word more frequently and consequently making note of it – the coincidences can seem uncanny.

I recently experienced a pertinent example of the Baader-Meinhof effect myself. After much pestering from my kids, I sat down with them to watch the musical biopic *The Greatest Showman*. The movie tells the story of the life of P. T. Barnum and the fates of the performers in the Barnum & Bailey Circus he founds. Although I didn't recall having heard of Barnum before, after watching the dramatic events of his life unfold over an hour and a half, he was at the forefront of my mind. When researching for this chapter only a week or so later, I, of course, came across Barnum statements (the general-purpose statements we met earlier in the chapter, designed to elicit the Forer effect) and made the inevitable connection. As I was already aware of the Baader-Meinhof effect, I saw this for what it was – an amusing coincidence, rather than a portentous omen that I should run off and join the circus.

A few weeks later, after debating the relative merits of Leonardo DiCaprio films with a friend (I don't believe there are many bad ones), he suggested I rewatch *Gangs of New York*, and there Barnum was again – a fringe character I never took in the first time I saw the film as a student. *The Greatest Showman* wasn't the first time I'd met Barnum at all – just the first time he'd stuck in my mind so strongly (as the film's main character) that I remembered it when I came across him again shortly afterwards. Similarly, although I don't recall meeting the idea of synchronicity before, once Paula introduced me to it and suggested I look out for it, I felt sure I would encounter it again in the following days. As expected, in the course of writing this chapter, I have come across the idea independently several times.

Gone fishing

In order to direct our conversation away from the easy vagaries that Paula has been pushing at me so far and to move the conversation into the realm of the concrete, I decide to divulge a little bit more about myself. I tell her that I am writing a book (although I purposefully fail to mention that this sitting is part of my research for it) and that I have just published another (my first book, *The Maths of Life and Death*). I am careful not to give any details away about the subject matter of either and, instead, I ask Paula how she thinks they will do. She asks me to draw another randomising set of tarot cards, which I carefully do, trying to give the appearance of picking 'more randomly' this time. I slide them over the table and Paula consults them in silence for a moment before launching into a *fishing expedition*.

Playing the odds, she assumes that I conform to the stereotype that captures most aspiring authors: the unknown dreamer, desperately trying to convince the world of the importance of their first novel. She tacitly assumes that I'm writing fiction and makes the intentionally vague suggestion that my books will 'immerse readers in a different world'. Of course, any author, even of non-fiction, wants to believe they are transporting their readers out of their daily lives into another more inspiring place, so I nod noncommittally.

'OK' I say, which Paula takes as encouragement that she is on the right track. She follows up by telling me that the first book will do quite well, but that the second will be a best-seller, which of course, despite my scepticism, I am not disappointed to hear and would love to believe. When I push her for more specifics, however, the fishing trip starts to go awry. She suggests that one of the characters in my new book will be inspired by one of my children. A reasonable suggestion for a novel, but not for a popular-science book. She goes on to suggest I might teach English literature based on the success she has predicted for my novels, and compares the books I will write to the fantasy series of vampire novels by Anne Rice, at which point I am finding it hard to nod in agreement anymore, so I try to change the subject.

Fishing expeditions are another classic tool for the clairvoyant to gather information and to appear to deliver nuggets of seemingly unknowable information to the sitter. They usually start off with an educated guess – the bait – playing the odds by working off something that a sitter has already mentioned or some aspect of their appearance. It might be that the sitter is wearing a wedding ring or that they are in the age bracket where a parent or grandparent may have passed away. Paula, for example, weaves one of the few personal details I have divulged (that I have children) into her predictions for my books.

Even without personal information there are common lures that can be used to give the psychic a hit. When pretending to communicate with the dead, for example, many psychics will fish for a name by saying something like 'Does a man with a J or a G name mean anything?' This is simply playing the odds. For the last 150 years, J has been the most popular first initial for male children born in the United States, accounting for between 15 and 20 per cent of all given names' initials. Combined with the less popular G, the percentage is consistently above 20 per cent. If you can think of ten male relatives in your family, then, providing their names are independent of each other (i.e. that there isn't a proclivity for a particular name or letter in your family) the chances are nearly 90 per cent that at least one of them will have a name beginning with a J or a G, providing the psychic with a hit. For me, both my uncles Jeremy and Gerald and my brother Geoff fit the bill.

When you consider that some psychics work in group settings with multiple people in the room, the chances increase even further. In the UK or the US with a room of thirty people each thinking of just two different male relatives the chances of a hit are over 99.99 per cent. If the crowd is big enough, the medium might even practise a technique known as *shotgunning* – rapidly listing some of the more popular J or G names: 'Is it a John, a Jack, a Jason, a James, a Joe, a Jerry? Does that make sense to anyone here?' If they get a hit on one of these names, then the wrong answers they listed are quickly forgotten by the audience. If someone in the audience supplies a J or a G name that isn't one of the listed ones, the psychic can still make it seem like they were homing in on the right answer, but just hadn't had the chance to put their finger on the right name yet. Even something as common as a correct initial can convince people who want to believe. They forget to question why the loved one the psychic is allegedly communing with can only remember their own first initial and not their whole name. Obviously, they've been put through from the spirit world on a bad line.

Throwing out a seemingly specific, but actually vague and generally applicable, statement and allowing the audience to fill in the gaps is key to the psychic guessing game. For example, they will often try to bolster their credibility by quantifying their predictions with an apparently specific number, perhaps by saying something like 'I can see four people in the family'. They might start by venturing that the number corresponds to the number of siblings of the sitter. If they only had three siblings, the psychic will remind them not to forget themselves, bringing the number up to four. If it was just three children in the family, then the psychic will include both parents to make four other members of the family (excluding the sitter), and if the sitter has only one sibling, then the total family size of four will suffice. If the sitter is an only child, then the medium might even venture that the mother had a miscarriage to make up the numbers. If correct, this has the double impact of proving the psychic right and striking an emotional chord with the sitter, seeming to further bolster the psychic's knowledge of those who have 'passed to the other side'. Since as many as one in four pregnancies ends in miscarriage, finding only children whose mothers miscarried is not that unlikely.

Of course, if the number four doesn't apply directly to the sitter's family, then the psychic can ask them to search for it in their partner's or their parents' families or the family of the deceased loved one they are hoping to communicate with. Psychics rely on the willingness of the people for whom they are reading to increase the odds, to find the connections for them and to forget the missteps they made in shotgunning to the answer.

A bad trip down memory lane

Shotgunning relies, in part, on the *von Restorff effect* – one of a whole family of ingrained prejudices known as *memory biases*. As the name suggests, these cognitive deficiencies block or alter the recall of memories – a disposition which, in a sitter, tends to benefit the psychic. In 1933, psychologist Hedwig von Restorff discovered the tendency of participants in her experiments to remember an unusual item on a list of otherwise similar objects.[20] As an example of just how potent the effect is, read the following shopping list once. Then, while looking away, see how many items you can remember.

Bananas, oranges, pears, grapefruit, *giraffes*, grapes, lemons, tangerines, apples.

Now close your eyes and reel off as many items on the list as you can. It's unlikely that you'll remember all the items, but I'd be willing to bet that you remembered the giraffes. Not only did the formatting make 'giraffes' stand out from the list, but the contextual incongruity with the other objects attracts a disproportionate amount of your attention to that single item. The distraction provided by the distinct item can serve to lower the total number of objects recalled in comparison to a list in which all the items are in the same category. For the same reason, a registered hit among a psychic's shotgunned list of otherwise unrecognised names attracts a disproportionately large weight in memory, leaving less space for the other items.

Perhaps the most pertinent memory bias that benefits psychics is the confirmation bias that afflicts the majority of their clients (sceptical authors excluded, to some extent) – the ones who really want to believe in their powers. These sitters tend to recall, primarily, those psychic utterances which agree with their original expectations (the accurate epistemic reconstruction of personal information, which they believe the psychic could have no way of knowing) – often ignoring the times when the psychic gets it wrong. Selective memory acts in a complementary manner for the psychic's future-facing aleatoric predictions. If predictions are scattergunned at a sitter fast enough – so fast that they can't remember them all – it's likely that only those forecasts that bear a resemblance to something that actually comes to pass will be recalled.

This parallels the experience many of us will have had of recalling a dream only later in the day, when something happens to trigger a memory of it. This, of course, doesn't imply that the dream was in some way prophetic. Rather, it suggests that if we hadn't had the triggering experience, we wouldn't have recalled the dream or memory at all. In the same way, believers remember only the few predictions that seem to come true and forget many that don't – highlighting the hits and shrouding the misses.

The final icing on the memory cake for the psychic is *hindsight bias* – the distortion of our memories in light of the knowledge of later events. This can have the effect of making originally vague predictions seem to match subsequent events, as only the pertinent details are recalled and simultaneously remoulded to agree with what actually happened. Among the most notable dependents on hindsight bias are the disciples of Nostradamus. In his book *Les Prophéties*, the sixteenth-century French seer wrote a collection of 942 vague and metaphor-laden four-line poems (quatrains) which supposedly predicted the future. The following lines, frequently presented by his modern-day followers as proof of his vision, allegedly foretell the demise of the 1986 *Challenger* shuttle which broke up shortly after take-off:

> *From the human flock nine will be sent away,*
> *Separated from judgment and counsel.*
> *Their fate will be sealed on departure.*
> *Kappa, Theta, Lambda the banished dead err.*

In support of their claim, believers note that the company that man-
ufactured the defective part which led to the disaster was called
Thiokol – which almost looks like an amalgam of the Romanised
versions ('k', 'th' and 'l') of the Greek letters kappa, theta and lambda
from the last line of the quatrain, if you squint. The fact that seven
astronauts died and not nine – quite a big discrepancy, you might
have thought – is conveniently swept under the rug.

Notably, not one of Nostradamus' 942 'predictions' has ever been
used to predict a specific event *before* it occurred. They have only
ever been invoked retroactively in a ploy referred to as *postdiction*
or *prediction after the fact*. To be blunt about Nostradamus' abilities,
a forecast which can only be connected to the event it purports
to predict after that event has happened is about as useful as a
chocolate teapot.

The vanishing negative

There is evidence to suggest that the accuracy of the memories we
are able to recall is also influenced by heightened emotion. Indeed,
emotion can lead people to accept statements that they most desire to
hear, even if those statements are logically inconsistent – an example
of what psychologists call *motivated reasoning*. People who have
recently lost a loved one are often in such a heightened emotional
state. Many mourners, perhaps unable or unwilling to accept that
their loved one has passed beyond contact, are inspired to visit a psychic or medium. The grief that accompanies the death of a close friend
or relative is a painful process and it is completely understandable that
sufferers will be highly motivated to find and retain information that

comforts them. Of course, this does leave grieving clients, desperate to contact a recently lost loved one, significantly more suggestible to the psychic's guessed intimations about the deceased than they might otherwise be.

As I sense my time in Spiritualism running out (by checking the time), I resolve to test Paula's ability to find out about my own lost loved one – my father. (I should disclose here that my father is still very much alive, but I am interested to see whether Paula is able to figure this out or just carry on making her unsupported predictions regardless). So I ask her if there are 'any messages for me from the other side'. Paula dampens my expectations by caveating that she is just 'developing mediumship at the moment'.

'Is there anyone in particular you want me to contact?' she enquires.

'I'm interested in hearing from my dad who passed a long time ago,' I tell her.

'What was your father's name?' she asks me.

'Tim,' I reply.

After a protracted pause in which Paula closes her eyes and seems to be concentrating quite hard at looking relaxed, she comes back to me.

'I've got a male,' she informs me. 'He wasn't a tall man, was he?'

'No,' I answer, 'he was shorter than me and I'm not exactly huge,' I laugh, expecting her to backtrack.

'No, I didn't think so,' she parries, turning the original implication of her prediction on its head. She carries on, 'I see him standing in front of me, but for one reason or another he's feeling quite shy . . .' Now my dad is anything but shy; he is one of the most outgoing and vivacious men I know – the life and soul of any party. I wonder how she will try to wriggle out of this one. When I fail to show any recognition of her description in my face – no telltale smile of acknowledgment, no subtle nod of the head – she quickly picks up that she has it wrong and continues '. . . which is strange because he's usually so outgoing'. I can't help but nod in agreement and admire her skill.

These two about-turns are examples of much-practised psychic sleights of tongue known as *ex post facto declaratives* – statements which can be interpreted or reinterpreted after the fact. The first, which Paula uses to guess at my dad's height, is an example of the *vanishing negative*. The technique works using a construction known as a *negative tag question*, in which a positive question is tagged on to a negative statement, making the questioner's intent potentially ambiguous. It's a common ploy many of us will use in order not to cause offence to someone whose views we are not quite sure of. For example, the tag question 'You don't believe in psychics, do you?' might be answered, 'Yes, I do' – in which case a placating response might be, 'Oh yes, I thought you did'. Alternatively, if the answer comes back, 'No, of course not', then an appropriate reply might be, 'No, I knew you wouldn't believe in that mumbo-jumbo'. In the same way, the vanishing negative allows the psychic to discover important information about their sitter, while appearing to have known it all along.

The second reversal is an example of a *punctuated rainbow ruse*, giving one polarised aspect of a personality statement, then, after reading the non-verbal response cues, quickly reversing the statement if there is no clear hit. This two-part trick can be more effective than the basic rainbow ruse, as it allows the psychic to glean information, rather than simply scoring them a hit with the sitter. Additionally, on the occasions in which the first part of the ruse is correct, the psychic never has recourse to use the other half of the personality trait. The 'direct hit' seems more impressive to the sitter than a simple Barnum statement.

Warm reading

Paula tries again to divine some specific detail relating to my dad. This time, she attempts to predict how he died. 'He keeps telling me that he passed due to a problem in the chest region,' she guesses, waving her hand over her torso, from her neck down to her waist. Of course, the region of the body Paula has indicated with her gesture includes almost all the major organs – the liver, the stomach, the intestines, the pancreas and, of course, the heart and lungs. The bottom line is that, in the end, everyone stops breathing and their hearts stop beating. These are the ultimate markers of death, so a prediction of problems in the chest region will always be assented to by someone who wants to believe enough.

Most of the tools Paula has tried out on me this evening (the rainbow ruse, the vanishing negative, fishing, shotgunning, bequeathing psychic credit, etc.) could be classed as 'cold-reading' techniques, relying on reading my body language, appearance and reactions to extract information from me. But this last ruse is a catch-all designed to give a hit in almost any circumstance, much like a Barnum statement. The use of such generic statements is known as *warm reading*. The techniques are supposed to portray the reader as psychically intuitive, whereas, in reality, the intimations are carefully crafted to give a hit in the vast majority of cases, irrespective of extrasensory perception.

Clearly, since my dad is still alive, responding truthfully to Paula's suggestion about his cause of death presents something of a problem to me. My easiest option to clear this hurdle is simply to nod and tell her he died of a heart attack – another acknowledged hit for Paula. It shouldn't be underestimated, especially in group-reading situations, how often the subject of a psychic's attention will agree (or at least not actively disagree) with their statements out of a fear of being socially awkward.

The last few minutes of my reading with Paula are taken up with several less successful guesses about my dad, including him having worn a flat cap and having a connection with coal mining (these last two presumably informed by the vestiges of my northern accent that she has picked up on) and some general platitudes about how 'he is around me a lot' and 'that there is a lot of love coming from him', which present no possibility of being challenged: which spirit-seeker wouldn't want to hear that of a lost loved one?

Hot reading

Although evidently well versed in the psychic staples of cold and warm reading, it's clear from her low rate of successful predictions this evening that Paula has not gone as far as to delve into the murky waters of *hot reading*. To prepare for a hot reading, a psychic actively investigates prospective sitters beforehand in order to access the information they would be expected to arrive at by supernatural means. Traditionally, this has relied on the reader looking up their victims in the phone book, pretending to be a door-to-door seller or a missionary in the hope of striking up a conversation, exchanging information with other local mediums and even visiting cemeteries to look up the names of deceased loved ones on gravestones.

Before the age of the internet, hot reading usually required a lot of preparation and its practice, therefore, was typically restricted to well-known psychics reading for large audiences and making enough money to employ someone to do the preparation for them. In some cases, renowned psychics have even been known to employ stooge audience members who mingle with the real audience in the lobby before the show. It is their job to carefully select victims to grill subtly for information, which they then relate to their on-stage colleagues. Undercover journalists have reported that, for his TV shows, a celebrated American medium even speaks to some audience members himself before filming begins in order to extract information from them. Once the cameras are rolling, he then revisits those same people, using the information he gathered and appearing (to the TV audience at least) to give extraordinarily accurate readings. The cold readings celebrity psychics might throw in to keep the live audience members onside tend (unsurprisingly) to be less successful and typically end up on the cutting-room floor. Predictably, given everything that is alleged to go on behind the scenes, it has been reported that this medium also makes all his audience members sign extensive release forms, preventing them from disclosing almost anything that happens during the filming.

The advent of the internet has made hot reading significantly easier. Armed with the powerful, but easily accessible combination of the show's location and a list of attendees' names, Facebook and other social media platforms provide hot readers with unprecedented insights into the private lives of potential audience members. Fortunately, the internet has also made it easier for sceptical vigilantes to land hot-reading psychics in hot water.

Susan Gerbic and Mark Edward are one such pair of activists. In 2017, they assembled a crack squad of debunkers to expose psychic medium Thomas John. In the run-up to one of John's shows, their team created fake Facebook profiles for the pair under the aliases of married couple Susanna and Mark Wilson. They also populated a number of other fake profiles whose aliases would interact with the Wilsons' profiles, reminiscing about significant life events and dropping names of made-up relatives for John to pick up on, while leaving Gerbic and Edward completely out of the loop. The fake Facebook friends even tagged John in some of the posts detailing how excited Susanna was to be going to John's show and her hopes to get in contact with her recently deceased (and entirely fictional) twin brother, Andy. Mark's fake profile detailed his desire to be reunited with the spirit of his (also fictional) father who had died some years earlier from heart disease. The existence of imagined deceased relatives were the only details that the team shared with Gerbic and Edward before the show.

On the day of the reading, Gerbic and Edward, in character as Mr and Mrs Wilson, sat in the VIP seats in the third row of the audience in the hope that they would be called upon and subjected to John's 'extrasensory powers'. Three or four readings in, John began with 'I'm getting someone's twin'. Gerbic duly raised her hand and was invited into the spotlight. On stage, John proceeded to reel off the details about Andy's pancreatic cancer which Gerbic acknowledged and responded to with the right degree of fake, but sincere-looking emotion. Then John started to deliver details that the Wilsons' fake Facebook friends had shared online, but that Gerbic and Edward were not privy to. John wanted to know why he kept being given the name Steve. Gerbic ummed and aahhed before guessing that Steve was a close friend of Andy's (it was actually supposed to be Mark's father's name).

'Who is Buddy?' asked John. Gerbic told John uncertainly that it was a nickname for both her brother and father (when in fact it was the name of the dog that the sceptic activists had invented for Gerbic). No doubt John was confused that Gerbic didn't know the name of her own beloved pet, or that Mark couldn't remember the name of the father whose spirit he had been hoping to hear from. Nevertheless, by blundering through these parts of the readings, Gerbic and Edward had demonstrated conclusively that they had no idea of the faked details – details that John had so clearly harvested directly from Facebook. By being rigorous about blinding themselves to the falsified information, there was no way, after they exposed him, that John could claim to have read it directly from their minds – a common get-out for hot-reading psychics – as Gerbic and Edward clearly didn't even know the information themselves.

Two years after the sting was completed and the evidence had been painstakingly compiled and documented, the *New York Times* published a detailed exposé of John's hot reading, based on Gerbic and Edward's operation. The story went viral, leaving John's reputation and credibility in tatters. Despite the clear and obvious evidence to the contrary, John still claims that he is not a hot reader: 'NO I do not Google people. NO I do not research people. NO I do not go onto people's obituaries. I do not go onto Ancestry.com.'

In response to his claims, ever the consummate professionals, Gerbic and her team trawled through the online webinars John has for sale on his website to unearth more evidence of his psychic fraud. In screenshots captured from one such video, John's Google history is (unintentionally) visible and includes searches for obituaries of several individuals, as well as evidence of searches from Intelius.com. As it boasts on its website, 'Whether you want to reunite with your college roommate or learn more about the person your daughter is dating, Intelius is your go-to resource for finding people.' It all makes you wonder why a genuine psychic would be doing genealogy research when they have a direct hotline to the other side.

Harmless fun?

For me, going to see a psychic was a bit of fun, allowing me to experience, at close quarters, the gimmicks that have been keeping mediums, soothsayers, oracles and seers in business for thousands of years. But for many, a visit to a psychic is a desperate staging post in their downward spiralling journey of grief. For some, almost no price is too high to pay in exchange for the answers they seek. During particularly high-profile tragedies, like murders or missing-persons investigations, it is common for hordes of self-styled 'psychic detectives' to seek out, unsolicited, those who are in these emotionally vulnerable states.

One such high-profile case was the disappearance, in October 1989, of eleven-year-old schoolboy Jacob Wetterling. Jacob was abducted by a masked man wielding a gun, while cycling home from a video-rental store in St Joseph, Minnesota. His ten-year-old brother and eleven-year-old friend were the only witnesses. In the days that followed, the most crucial period in a missing-persons' case, police wasted vital time following up information from psychics. In one typically misguided diversion, less than a month after the abduction, a combined force – including the FBI, Iowa state police, local officers and sheriff's deputies from four counties – spent two full days searching farmhouses and sheds along a 25-mile stretch of road in Iowa, based on a tip from a New York psychic. They found nothing. While investigators were busy following this far-flung lead, they still hadn't interviewed all the neighbours who lived on the cul-de-sac where Jacob went missing. Nor had they talked to one of the prime suspects in the case, Danny Heinrich, who was already implicated in a similar abduction in a nearby town just nine months earlier.

In the years that followed, as the case remained unresolved, more and more psychics came out of the woodwork. Many of them requested some of Jacob's toys or clothes to help them in their 'inquiries'. Desperate to believe they could help, Jacob's father, Jerry, dutifully packaged up and sent these items off, many never to be seen again. Other mediums would call in the evenings and, feeling unable to bypass even a single avenue of investigation, no matter how tenuous the connection, Jerry would answer, often hearing them out late into the night. The time burden, coupled with Jerry's willingness to believe their phony proclamations, slowly drove a wedge between him and Jacob's mother, Patty.

Infamous psychic Sylvia Browne even called Jerry to tell him that Jacob had been abducted by two men from Illinois. Browne was no stranger to weighing in on missing-persons cases. In one particularly grievous blunder she told Louwana Miller, the mother of missing teenager Amanda Berry, that her daughter was dead, only for Berry to turn up, very much alive, after being held captive for ten years. Tragically, Miller died long before Berry escaped, Browne's intervention crushing her last remaining hope that her daughter would ever come home. In common with most of Browne's other missing-persons predictions, she was wrong about Jacob Wetterling. In 2016, twenty-seven years after Jacob's abduction, the man who should have been the chief suspect, Danny Heinrich, eventually admitted to having abducted, sexually assaulted and murdered Jacob, all on the same night, as well as having gone on to commit numerous other sexual offences in the intervening years. Who knows how things might have played out had police focused on compelling local evidence, rather than spending their resources following the dead-end trails fed to them by purblind seers?

Only in high-profile cases is it common for psychics to make unsolicited approaches to their victims. However, many people who have lost loved ones will actively seek the comforting words of a psychic themselves. The best possible outcome from such a visit is that the seeker goes away believing that they have genuinely heard a reassuring message from a loved one or, having received some harmless advice, deciding to leave it at that. Unfortunately, there are some unscrupulous practitioners out there who set out with malign intent – to commit psychic fraud. To all but the most credulous, the phrase 'psychic fraud' seems tautological. Surely people seeking psychic advice are being deceived into parting with their money in return for something with no intrinsic value – almost the very definition of fraud. But the legal difference makes itself plain when mediums seek to extort money from their patently vulnerable clients.

Fortunately, there are people out there who are fighting back against these clairvoyant con artists. Perhaps best known among these vigilantes is private investigator and scourge of psychics, Bob Nygaard. Nygaard has a long track record of bringing fraudulent psychics to justice. His first case was that of a Miami doctor who'd been conned out of $12,000 by psychic Gina Marie Marks. Marks had told the doctor that her anxiety problems were caused by a disagreeable colleague who'd buried a piece of meat in order to put a curse on her – a story so bizarre that the client believed it couldn't have been made up. According to Marks, the only way to cure the curse was to perform a mystic ritual involving stroking an egg while burning special candles – and for the doctor to hand over thousands of dollars to Marks, so that the egg could be 'cleansed'. During his investigations, Nygaard found four of Marks' other victims who collectively had been defrauded of $340,000. After a decade-long investigation, he'd eventually gathered enough evidence to send Marks to jail for six years.

Another case that Nygaard investigated, notable for the depth and sheer implausibility of the psychic's deceits, was that of thirty-two-year-old victim Niall Rice. Rice went to see Pricilla Delmaro, a New York-based psychic, when he found out that Michelle, the woman he loved, didn't share his feelings. To try to win Michelle over, Delmaro had Rice buy all manner of expensive gifts, including a $30,000 gold-plated Rolex. But these gifts weren't to give to Michelle to win her favour. Instead, the Rolex was used as part of an elaborate ceremony to 'turn back time' and 'cleanse his past'. Delmaro even convinced Rice to part with $80,000 to buy an imaginary 80-mile gold bridge in the spirit realm, which would be used to distract an evil spirit. When it turned out that Michelle had died of a drug overdose, Delmaro offered (for a fee) to reincarnate her in the body of a thirty-one-year-old woman. Rice's journey to Los Angeles to meet the 'new Michelle', who didn't seem to be much like the original, led him to the suspicion that, in his words, 'Delmaro wasn't everything she was purporting to be'.

By this point, Rice had sold his apartment, was over half a million dollars out of pocket and, in desperation, called on Nygaard's services.

Despite Rice's gullibility and incaution (even sleeping with Delmaro at one point), Nygaard was able to secure a conviction.

Lessons learned

The subtle psychological manipulations that modern-day psychics use to beguile their victims are the same tricks that oracles, sooth-sayers and seers the world over have been using for years. When Croesus, King of the Lydians, consulted the Oracle at Delphi about whether to act against the increasing power of the Persian Empire in his native Anatolia, he was told in no uncertain terms that 'If you cross the river, you will destroy a great empire'. Believing the prophecy to be favourable, he duly launched his campaign against the Persians in 547 BCE and a great empire was destroyed – his own. The Oracle, of course, was vindicated by the whole affair, with some commenta-tors suggesting, postdictively, that this is what the Oracle's statement had meant all along. Of course, just like the rainbow-rusing mediums of today, the Oracle's base-covering prediction, that there would be a winner in a battle, was never likely to be wrong. You don't get to the top of the most powerful and well-respected forecasting institution of antiquity by risking mistaken predictions.

When a Greek general went to consult the Oracle at Dodona about his fortune in an upcoming battle, the Oracle is alleged to have replied, 'You will go, you will return never in the war will you perish'. This is a beautiful early example of an *ex-post-facto statement* – in the same category as the psychics' vanishing-negative trope. The sentence is deliberately ambiguous with two diametrically opposed meanings that can be interpreted, after the fact, to give the appearance of a correct prediction. If the general died, it was meant to be understood as, 'You will go, you will return never, in the war will you perish'; but if the general returned triumphant, it was always, 'You will go, you will return, never in the war will you perish'. The Latin phrase *ibis redibis* is used in a legal context to refer to a confusing or ambiguous statement and comes directly from the Latin translation of the Oracle's prediction: '*Ibis redibis nunquam per bella peribis*'.

Then, of course, there was Nostradamus' shotgunning. Contained within his almost 4000 lines of vague predictions are bound to be a few sentences which bear a resemblance to notable events in history. In some cases, though, Nostradamus' predictions were so ambiguous that the same quatrain has been interpreted as predicting very different events. It seems obvious that if you make enough of these equivocal predictions, some of them are bound to come true when squinted at through the lens of hindsight. And the ones that don't? They're just moved to the pile of predictions yet to come true.

And this is precisely the point. The majority of predictions from antiquity are either so non-specific that no one can really tell if they have so far come true or were incorrect but unremarkable, and consequently they vanished without trace because they didn't make very good stories. No one remembers the thousands of times the priests, haruspices and 'wisemen' got it wrong, because their false steps were eclipsed in the records by the one or two chance occasions when they seemed to get it right.

As my session with Paula draws to a close, I thank her for her efforts. In spite of the fact that I don't feel she has brought me much enlightenment, I appreciate that she is good at what she does. She couldn't get through half an hour of speculating about the life of a total stranger – with the expectation that she will reveal some fundamental truths – if she wasn't.

In return, Paula leaves me with the reassuring message: 'Good luck with the book; you'll be fine,' which resonates with me more strongly than she will realise as I step out on to the street to find somewhere to sit down and gather my thoughts. I alight upon a bench just around the corner and try to recall the items in the psychic toolbox that Paula has brought to bear on me today, as well as the cognitive biases that we all possess, which can render us unwitting accomplices in these scenarios. If we are not to be taken unawares, then we must learn to recognise these mental rules of thumb that, as a species, we have developed over millennia of human evolution: the Pollyanna principle; the memory biases; the magical thinking; and the recency effect – our own gut feelings – when they come to our 'aid'. If we don't, then we leave ourselves vulnerable to exploitation by psychics wishing to bedazzle us, to profit from our insecurities or simply to avoid being exposed. As we'll see in the next chapter, however, we don't need an outside agent to deliberately lead us astray in order to find ourselves deceived. We're quite capable of doing that for ourselves.

2

EXPECTING THE EVERYDAY EXTRAORDINARY

'Ha,' I laughed, 'you're such a nerd. I can't believe you print out a special bookmark for every book you read.'

My brother, Geoff, had just arrived at the house that our family was staying in that summer holiday back in 2009. After hugs and pleasantries, I picked up the book he had been reading on his journey and that's when I spotted it. Poking out of the top of Paul Auster's *New York Trilogy* was a piece of card with 'P AUST' printed near the top edge.

'What are you talking about?' he asked me. 'I don't do that.'

'Whatever,' I teased. 'Somebody else does it for you then, do they?'

'What are you on about?' he laughed, somewhat bemused. He genuinely seemed to have no idea what I was talking about. So I held up the book to show him.

'This, see?' I chided.

'What?' he asked, uncomprehending. 'That's just my train ticket.'

Incredulous at his denial, I whipped the bookmark out of the book and examined the piece of card. Indeed, it was just a train ticket. Well, not just a train ticket, but a train ticket that held that unmistakable abbreviation, P AUST. I couldn't make sense of it.

Examining the ticket further, I saw that it was for the penultimate leg of Geoff's journey between Paris and Limoges. Suddenly, everything made sense: what station do you have to embark from to reach Limoges from Paris? I'd made the same trip from this station only the week before: Paris Austerlitz, or P AUST to train aficionados.

Reflecting on this over the next couple of days, I started to doubt how much of a coincidence it really was. The probability of such a thing happening by chance seemed far too small. Geoff would have been sent his tickets by post well in advance. It would have been just his sort of ruse to have noticed the abbreviation and to have brought along the Auster book to give us all something to talk about. Perhaps he had planned the whole thing?

I asked to see the ticket again. Looking at it more closely, I was surprised at what I found. Nowhere on the original printed ticket did an abbreviation of the station name appear. It was written Paris Austerlitz in full throughout. The P AUST stamped vertically on the left-hand edge of the ticket was clearly a later addition, but not from Geoff, or at least not intentionally.

In France, when you take a train, you are first supposed to voluntarily 'compost' or validate your ticket (a system originally designed to stop opportunistic travellers from using open-ended tickets more than once). You put your ticket into a machine on the platform and it comes out with a printed stamp or 'obliteration code' unique to your embarkation station. Only when Geoff composted his ticket on the platform at Austerlitz was the P AUST franking added. Pre-planning this seemed extremely unlikely.

*

When coincidences appear without warning they can send us sprawling in search of the reasons behind them. As we saw in the previous chapter, psychics label common coincidences as synchronicity and exploit this idea to win buy-in from their sitters, endowing them with psychic credit. In our everyday hunt to draw out meaning from the events that befall us, we are often too quick to infer causation when there may not be any. My first thought upon seeing the train ticket in the Auster book was to assume that my brother had developed a niche penchant for bespoke bookmarks – in hindsight, an unlikely proclivity. Even after I discovered the nature of the occurrence, I still ascribed the event to his mischievous intervention, until I could convince myself that, beyond doubt, the most likely cause was simple raw chance.

Almost everyone, even the most rational among us, will, at some point, have been taken by surprise by a seemingly impossible coincidence. The effect may have been significant enough to lull us temporarily into believing something that, with reasoned thought, is extremely unlikely or perhaps even impossible. Many of the characters claiming supernatural clairvoyance or psychic ability in the previous chapter may have experienced just such a powerful coincidence that changed or confirmed their mystical world view to such a degree that they were never able to reason their way beyond it.

Quite naturally, most of us have an egocentric take on the world. Understandably, we think about the probability of chance events from our own point of view first and foremost. It is hard for us to overcome this long-honed intuition and to reconcile ourselves with the fact that some seemingly improbable events are, in fact, exceedingly likely, if not completely certain.

In raffle lotteries, like the UK's (now discontinued) Millionaire raffle, one winning ticket is selected from among hundreds of thousands of entries each week. When you don't win, you have no problem accepting that the chances of your winning were extremely low. However, if your numbers were to come up, it might be incredibly tempting to attribute your luck to fate or some higher agency marking the cards, even though someone somewhere is guaranteed to win on each draw. Quotes from winners of such draws are littered with 'signs from above' or 'messages from relatives on the other side'. If your ticket is a close match to the jackpot-winning numbers, but not quite right, the idea that you are the butt of some cosmic joke is hard to shake, despite the fact that you were no closer to the main prize than any of the other numbers that didn't win. You either win or you don't.

Reconciling our individual lived experience of luck with the cold, rational science of large-scale probability isn't always easy. We look for signal and significance in places where only static prevails, and sometimes we find it.

Blinded by *coinscience*

Science is an area in which spotting coincidences can be extremely useful. In 1912, for example, German climatologist Alfred Wegener noticed the apparently strange coincidence that the coastline of West Africa and the eastern coastline of South America seemed to fit together like pieces in a jigsaw puzzle. Despite the prevailing opinion at the time, that the enormous land masses which comprised the continents were just too big to move, Wegener proposed the only theory that reconciled his observations:[21] continental drift suggested that the land masses weren't rooted in place but could, ever so slowly, change their relative positions on the surface of the earth.

When he published his theory in 1915,[22] he became a laughing stock. Geologists rejected his outlandish idea, citing a lack of mechanism for moving such enormous chunks of the earth's surface, dismissing the seemingly snug tessellation of the continents as pure coincidence. By the 1960s, however, the theory of plate tectonics[23] – the movement of the solid mantle and crust over the earth's surface – gave credence to Wegener's now widely accepted theories.

In 1815, English physician William Prout noticed that the atomic weights of the elements that chemist John Dalton had recently measured were roughly whole-number multiples of the atomic weight of hydrogen.[24] This led him to suggest that atoms of other elements would be amalgamations of various numbers of hydrogen atoms.[25] For example, roughly eight grams of oxygen were needed to combine with one gram of hydrogen to make water. Since we know that every water molecule (H_2O) contains two hydrogen atoms for every oxygen atom, by Prout's theory, an oxygen atom should (and does) weigh around sixteen times as much as a hydrogen atom. Based on similar, approximately whole-number ratios for the other elements, Prout suggested that hydrogen was the only truly fundamental particle (which he dubbed 'protyle') and that atoms of other elements were made up of different numbers of hydrogen atoms.

Later, more accurate experiments showed that the weights of atoms of other elements were *not* close to being whole-number multiples. Chlorine presented a particular problem.[26] The formation of hydrochloric acid (HCl), comprising one atom of chlorine and one of hydrogen, required roughly 35.45 grams of chlorine to react with 1 gram of hydrogen. This suggested that chlorine atoms have an average weight 35.45 times that of hydrogen, casting serious doubt on Prout's 'whole-number-ratio' hypothesis.

Indeed, as it transpired, Prout was not quite right. Atoms are, in fact, made up of protons, neutrons (which have almost exactly the same mass as protons) and electrons (which weigh around 2000 times less, meaning they make little difference in the calculations). There are also different versions of the same element called isotopes, which have the same number of protons, but different numbers of neutrons. Chlorine, for example, has two main isotopes – one with seventeen protons and eighteen neutrons and, therefore, a mass of roughly 35 times that of a hydrogen atom, and another with seventeen protons and twenty neutrons and a mass roughly 37 times that of a hydrogen atom. ^{35}Cl and ^{37}Cl occur in a ratio of roughly three to one, explaining why approximately 35.5 ($¾ \times 35 + ¼ \times 37$) grams of naturally occurring chlorine are needed to combine with 1 gram of hydrogen. Despite this nuance, when considered for each isotope, Prout's suggestion that the mass of other atoms would be roughly integer multiples of the mass of a hydrogen atom was correct and became known as the 'whole-number rule' for which Francis Aston, who articulated it in this form, was awarded the 1922 Nobel Prize in Chemistry.

Perhaps more importantly, by spotting a pattern in the disorderly measurements, Prout's insight served to stimulate the academic debate, which led to significant improvements in our understanding of atomic structure. When, over a hundred years later, Ernest Rutherford fired alpha particles at nitrogen atoms to displace hydrogen nuclei,[27] he conjectured that all atoms might be made up of these fundamental particles. He went on to name them protons, which, as well as deriving from the Greek word *protos* meaning 'first', satisfied Rutherford's additional intention of honouring Prout for his insightful conjecture.

Although coincidences can point the way to new scientific discoveries, they can also prove an obstacle to scientific progress when they appear to confirm an incorrect theory. In the early 1800s, German anatomist Johann Friedrich Meckel made just such a blunder. He was a believer in *scala naturae* (the ladder of nature), in which humans sit above all other animals in an ordered but static hierarchy. The simplest, most primitive life forms were supposed to sit on the lowest rungs of the ladder, while the most complex and advanced beings resided on the highest. His views were hardly surprising given that this 'great chain of being' was the predominant theory of the day. The now generally accepted theory of 'common descent' – that multiple species descend from a single ancestral population – was only in its infancy as an idea at the time.

Meckel employed *scala naturae* to come up with a conjecture about his area of speciality – embryonic development. In what he called recapitulation theory,[28] he posited that, as they developed, the embryos of higher-order animals (like mammals) progressed successively through forms which strongly resembled the 'less perfect' animals, like fish, amphibians and reptiles, on lower rungs of the ladder. One startling, but seemingly unlikely prediction of this theory was that, as humans progressed through the 'fish stage', their embryos would have gill slits.

As it happens, it was discovered in 1827 that human embryos really do have slits[29] that resemble gills at an early stage of development.

This extraordinary finding seemed to bear out Meckel's prediction and corroborate his recapitulation theory. So strong was the perceived evidence that the theory became widely accepted, and it wasn't until almost fifty years later, in the 1870s, that the recapitulation theory of development was finally put to bed for good as the idea of common descent started to take hold.[30] Common descent made it clear that, far from undergoing a 'fish stage' in the womb, the gill slits were a consequence of the fact that, sharing a common ancestor with fish, we also share much of their DNA and early developmental processes.

Patterns in the noise

In a world awash with data, modern-day scientists need to be even more careful that they don't misinterpret random coincidences as significant links. When answering scientific questions, we are often interested in determining whether one quantity varies with another. For example, we might want to understand whether the prevalence or absence of a particular environmental factor increases or decreases our chances of a given health impact.

In February 1992, Julie Larm's eldest son, Kevin, was diagnosed with acute lymphocytic leukaemia. The mother of five from Omaha, Nebraska, recalled, 'That very day I wanted to know what caused that cancer, because I was afraid for all my children'. After mobilising a group of parents of other childhood cancer patients to form the 'Omaha parents for the prevention of cancer' group, Julie and her team set about investigating the potential causes. They plotted all recent known cases of childhood cancer on a map of Omaha, and found some distinctive regions in which cases appeared to be clustered together. When they overlaid their map with a mesh representing electricity cables, they found that some of the areas containing cancer clusters were densely criss-crossed with powerlines. They also discovered that there were at least eleven children living within a 1-mile radius of an electricity substation in Omaha who had been diagnosed with cancer within the previous seven years.

Throughout the 1980s and early 1990s, significant interest had developed in understanding whether living near powerlines increased the risk of developing cancer. By 1992, several well-renowned physicists had already weighed into the debate, pointing out that the strength of electromagnetic fields emitted by powerlines are hundreds of times weaker than the earth's magnetic field and thus unlikely to cause damage.

One imaginative idea pushed by the 'powerlines-cause-cancer' proponents was that the body's cells became 'entrained' to the frequency of the oscillating electromagnetic field. But physicists again calculated that any potential forces induced by powerlines on human cells would be thousands of times less than the strength of the fluctuations caused by the body's own heat. On top of this, biologists found it hard to explain how these tiny forces might induce cancer. In short, there were no plausible physical or biological mechanisms to connect powerlines to cancer.

However, as long as no conclusive trial had been conducted to empirically rule out the link, the Omaha parents' group continued to believe that overhead electricity cables were the cause of the cancer clusters they had identified. As we shall see, the most likely driver for the clusters was, in fact, pure unadulterated chance.

*

To understand why the parents' group were ultimately wide of the mark we need to know something about how humans deal with randomness (when they're not, as in the previous chapter, exploiting it as a blank page on which to compose their psychic messages). Unfortunately, as it turns out, when it comes to understanding random phenomena, our intuition often lets us down.

Before you read the caption, see if you can pick out the data set in Figure 2-1 (below), generated using truly uniform random numbers for the coordinates of the dots (i.e. for each point, independent of the others, the horizontal coordinate is equally likely to fall anywhere along the horizontal axis and the vertical coordinate is equally likely to fall anywhere along the vertical).

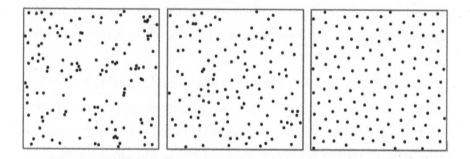

Figure 2-1: Three data sets, each with 132 points. One represents the position of the nests of Patagonian seabirds, another the position of ant colony nest sites and the third represents randomly generated coordinates. But which is which?

If you weren't sure and decided to plump for the central image, the chances are you're suffering from 'middle bias' – the tendency to exclude extreme options in favour of more central ones. Behavioural scientists have shown[31] that when choosing between *two* pricing-plan options, most people tend to choose the basic rather than the premium option, but that when a *third* ultra-premium option is introduced, the now 'mid-range' premium option becomes the most popular. It's worth questioning, when hedging your bets against the future with an insurance policy, whether the 'platinum option' really offers any tangible benefits or is just there to make the 'gold option' more palatable.

Similarly, educational psychologists have found[32] that students who genuinely don't know the answer to multiple-choice questions tend to favour the middle two options of four, or the centre-most option of five for their guesses. The same effect occurs when playing Battleships (where the tendency is to guess coordinates away from the edges disproportionately often when trying to sink an opponent's ship), when choosing items on a shelf or options from a computer drop-down menu,[33] and even when visiting the toilet (where a middle cubicle is up to 50 per cent more likely to be chosen than an outer one).[34]

As it happens, the truly randomly distributed points in Figure 2-1 are those in the left-most image. The middle image represents the position of ants' nests that, although distributed with some randomness, demonstrate a tendency to avoid being too close together in order not to overexploit the same resources. The territorial Patagonian seabirds' nesting sites, in the right-most image, exhibit an even more regular and well-spaced distribution, preferring not to be too near to their neighbours when rearing their young. The computer-generated points, distributed uniformly at random in the left-hand image, have no such qualms about their close proximity.

If you chose the wrong option, you are by no means alone. Quite apart from the possibility of middle bias, most of us tend to think of randomness as being 'well spaced'. The tight clustering of dots and the frequent wide gaps of the genuinely random distribution seem to contradict our inherent ideas of what randomness should look like.

It is precisely because of this long-honed cognitive bias that some of my recent research[35] has focused on developing metrics which are able to tell whether a spatial pattern is random or not, completely taking human perception out of the equation. Researchers in my group use these tools to determine whether cells in a developing embryo are more spaced out than we might expect and to characterise and better understand the beautiful striped patterns of zebrafish.[36]

*

Understanding that random patterns are not well spaced sheds some light on Omaha's supposed cancer clusters. Since cancer cases are not evenly spread across a country, but scattered randomly, even if no carcinogenic agent exists, randomly distributed cancer cases will fall into clusters purely by chance. What the Omaha parents' group had stumbled upon in detecting patterns in the noisy spread of cancer cases was likely an example of a logical mistake caused by randomness and known as the *Texas sharp-shooter fallacy*.

The fallacy gets its name from a story about a Texan cowboy who likes to head out to his barn after a few drinks for target practice. Invariably, the barn wall gets peppered with random bullet holes during the inebriated exercise, but purely by chance some of these are clustered together. One morning, the savvy 'sharp-shooter' gets out his paint cans and daubs a target around this cluster of holes to give the impression of accuracy to anyone who didn't see the process by which they were made and to draw attention away from the other more dispersed bullet holes.

The sharp-shooter fallacy occurs when a conclusion is drawn

based only on data consistent with a given hypothesis, ignoring data that don't support the proposed conclusion. The mistaken identifications of linked cases and the subsequent target-drawing conclusions aren't always done consciously. In some senses, the fallacy could be considered the bastard child of confirmation bias and hindsight bias that featured prominently in the previous chapter – homing in only on what we expect to see after the noisy data have been generated. The production of reality-TV shows is a classic example where we can see purposeful sharp-shooters at work. Film enough people for long enough and even the most mundane conversation snippets can be edited to paint a compellingly targeted narrative.

By painting a target around those cancer clusters which fell within areas with large numbers of powerlines, the Omaha parents had, perhaps unwittingly, committed the sharp-shooter fallacy. While understanding the fallacy doesn't necessarily rule out a link, it does suggest that the clusters of cancer cases in Omaha neighbourhoods, both heavy and light on powerlines, might be appearing without any overarching generative cause.

*

Randomness can leave our human brains poorly placed to make sensible deductions and, unfortunately for us, it is a part, to a greater or lesser degree, of many everyday situations we are faced with, from the arrival time of the next bus to the next song that is dealt to us by our music players when set to shuffle.

As a case in point, after noticing a disproportionate number of Steely Dan songs playing on his iPod shuffle, journalist Steven Levy questioned Steve Jobs directly about whether 'shuffle' was truly random. Jobs assured him that it was and even got an engineer on the phone to confirm it. A follow-up article Levy wrote in *Newsweek* garnered a huge response from readers having similar experiences, questioning, for example, how two Bob Dylan songs shuffled to play one after the other (from among the thousands of songs in their collections) could possibly be random.

We ascribe meaning too readily to the clustering that randomness produces and, consequently, we deduce that there is some generative force behind the pattern. We are hard-wired to do this. The 'evolutionary' argument holds that tens of thousands of years ago, if you were out hunting or gathering in the forest and you heard a rustle in the bushes, you'd be wise to play it safe and to run away as fast as you could. Maybe it was a predator out looking for their lunch and by running away you saved your skin. Probably, it was just the wind randomly whispering in the leaves and you ended up looking a little foolish – foolish, but alive and able to pass on your paranoid pattern-spotting genes to the next generation.

Nowadays, in the absence of the danger of predation, the main use to which our species puts this long-honed auditory skill is in alleging *backmasking* – the practice by which a reversed message is included in an audio recording. Despite the fact that it makes little sense, people have claimed to hear, 'Here's to my sweet Satan, the one whose little path would make me sad, whose power is Satan. He will give those with him 666. There was a little tool shed where he made us suffer, sad Satan,' when Led Zeppelin's 'Stairway to Heaven' is played backwards.

Despite denials by the band that they had deliberately encoded secret messages into their tracks, a 1982 session of the Consumer Protection Committee of the California State Assembly was played reversed segments of 'Stairway to Heaven' and asked to vote on a bill (which ultimately didn't pass) that would see mandatory warning labels placed on music containing 'dangerous' backward messages. Self-declared 'neuroscientist' William Yarroll testified to the committee that teenagers need only hear backmasked songs three times before the subliminal messages were 'stored as truth', turning them into disciples of the antichrist. Although Led Zeppelin were deemed the principal offenders, Yarroll alleged that he had found messages in the reversed music of other bands, including Queen and the Beatles.

The reversal of the repeated words 'number nine' in the Beatles 1968 track 'Revolution 9' sounds a little bit like the phrase 'turn me on, dead man'. Played backwards, John Lennon's mumblings at the end of the song 'I'm So Tired' from the same album sound like 'Paul is dead, man. Miss him, miss him, miss him'. These 'discoveries' lent credence to the 'Paul-is-dead' conspiracy theory – that Paul McCartney had died in November 1966 and was covertly replaced with a doppelganger. If you listen to any of these reversed pieces, you'll hear that the correspondence between the noise and the ascribed words is pretty tenuous – that listeners are finding patterns where there probably aren't any.

This auditory wishful thinking is just one example of the phenomenon known in the psychology literature as *pareidolia*, in which an observer interprets an ambiguous auditory or visual stimulus as something they are familiar with. This was the 'patternicity' phenomenon I described in the introduction, which allows my children to spot shapes in the clouds and is the reason why people think they see a man in the moon. Pareidolia is itself an example of the more general phenomenon of *apophenia*, in which people mistakenly perceive connections between and ascribe meaning to unrelated events or objects. Apophenia's misconstrued connections lead us to validate incorrect hypotheses and draw illogical conclusions. Consequently, the phenomenon lies at the root of many conspiracy theories – think, for example, of extraterrestrial seekers believing that any bright light in the sky is a UFO.

Apophenia sends us looking for the cause behind the effect when, in reality, there is none at all. When we hear two songs by the same artist back to back, we are too quick to cry foul in the belief that we have spotted a pattern, when in fact, these sorts of clusters are an inherent feature of randomness.

Eventually, the dissatisfaction caused by the clustering inherent to the iPod's genuinely random shuffle algorithm led Steve Jobs to implement the new 'Smart Shuffle' feature on the iPod, which meant that the next song played couldn't be too similar to the previous song, better conforming to our misconceived ideas of what randomness looks like. As Jobs himself surmised, 'We're making it less random to make it feel more random'.

*

Julie Larm's campaigning, based on her conviction that she had spotted patterns of cancer cases, took her and the other Omaha parents all the way to the Nebraska state health department. But when they got there, officials explained away their evidence as a simple clustering effect – a textbook case of apophenia. Just when they were losing hope that their theory would ever be believed by the authorities, a huge study from Sweden was published[37] which appeared to lend support to their hypotheses.

The Swedish study purported to show that children who were highly exposed to the electromagnetic fields generated by powerlines were nearly four times more likely to get leukaemia than unexposed children. The risk ratio of four (rate of cancer cases in those exposed to powerline electromagnetic fields divided by the rate in the control group) suggested an extremely strong effect and it seemed difficult to explain this result away as a fluke caused by noise in a small population, because the sample size was huge. The researchers in the study had been extremely diligent, enrolling everyone who lived, for at least a year, within 300 metres of Sweden's 220 or 400 kilovolt powerlines between 1960 and 1985. They also went to great lengths to calculate the strength of the electromagnetic field the patients were subjected to at the time of their cancer diagnosis and before. Surely a finding of this strength from this huge and seemingly unassailable study constituted incontrovertible proof that powerlines caused cancer – proof that even the physicists and biologists who had cast doubt on the mechanisms would have to accept.

As it turned out, there was a problem with the study – a problem so prevalent that it is one of the first lessons that students of public-health epidemiology are taught. Instead of deciding exactly what it was they wanted to look at before they set off on their investigation, the authors took a substantial range of measurements and made a huge number of comparisons. Their extensive, meticulously gathered and highly stratified data set, although impressive sounding, was to be their downfall. It allowed them to make not just one simple comparison between rates of cancer for people who lived near powerlines and those who didn't, but many. They were able to undertake similar analyses for large numbers of subsets of people: people who'd lived near powerlines two, five or ten years before diagnosis; people who'd been exposed to 0.1 or 0.2 or 0.3 microtesla field strengths; people who'd lived their whole lives in the vicinity of the powerlines or just part of them; people who lived in single apartment dwellings or people who lived in multiple-occupancy apartment blocks; adults or children – the list went on. In each of these subpopulations, the study checked for increased incidence of a range of different diseases, eventually leading to the calculation of over 800 risk ratios. Although on the surface this sounds extremely thorough, in reality, it is a fundamental scientific error known as the *multiple-comparisons fallacy* or the *look-elsewhere effect*.

When attempting to gather data on the variation of the environmental factors and compare it to the incidence of health issues, for all sorts of reasons, we typically find variation in the data. Fortunately, even when the data are noisy, there are many statistical tests designed to assess our confidence that a given relationship between factors exists. These tests often result in what's known as a *p*-value, where the *p* stands for probability. Roughly speaking, the *p*-value is the probability that results as extreme as the observed data might have been obtained if there were genuinely no relationship between the environmental factor and the health issue. The lower this *p*-value is, the more confident we can be that any correlation found between two factors is a true reflection of their relationship and not simply a result of chance. For example, a *p*-value of 0.05 suggests that, on average, such a relationship between two factors would only be observed by chance (if no relationship between the factors genuinely existed) once in every twenty repeats of the experiment (the probability 0.05 being the same as 1/20). The finding doesn't prove anything definitively. However, finding that the probability of seeing the data is low if there is really no relationship between the variables improves our confidence that such a relationship exists.

Usually, an acceptable value of *p*, known as the *significance level* of the study, should be specified before any evidence is gathered. If the value of *p* found is less than the significance level, then the relationship being tested is said to be *statistically significant*. Different areas of science demand different levels of confidence in their results. The more confident you want to be of your findings, the lower the value of the significance level you should specify. Using the 0.01 significance level, for example, means that even when there is no real effect, on average, for every hundred independent repeats of the experiment, the statistical test will suggest that there is. As the number of tests you do goes up, with a fixed significance level, the chance of finding a seemingly statistically significant result increases. One simple way to correct for this – known as the Bonferroni correction – is to divide your significance level by the number of different tests you are doing. The more tests you do, the harder it becomes for a result to be considered significant if there genuinely is no relationship. When multiple comparisons are applied deliberately to a data set and only the significant ones reported, this is known as data dredging, data fishing or p-hacking.

When performing a big study, if the proposed test fails to show a significant result, then, by stratifying the data, it is possible to 'look elsewhere' by testing a range of potential effects induced in different subsets of the data – specifically in the Swedish powerline study the associations between powerlines and a range of diseases in different subpopulations. In summarising their results, the authors of the study wrote: 'For brain tumours or for all childhood cancers together there was little support for an association'. But the authors went on to make over 800 comparisons, making the chances of finding a seemingly significant result, even at a high significance level (low-p value), extremely likely. They eventually found an elevated risk of getting myeloid leukaemia for children, living in one-family houses and experiencing exposures of over 0.3 microtesla. You can see for yourself from the multiple caveats in the previous sentence how many conditions there are, restricting both the specific disease and the subpopulation of the study's participants in which the result became significant.

In reality, because of the huge number of comparisons, we can't be confident that this niche association didn't occur purely by chance. Believing their conclusion to be significant, the researchers found themselves trying to suggest implausible reasons why only this specific subpopulation of individuals was affected and only in this specific way, even calling into doubt the validity of their own electromagnetic-field estimation techniques in multiple-occupancy apartments to justify the fact that the result was only significant in one-family homes. Some scientists, it seems, are just as vulnerable as the rest of us to making causal inferences when faced with random coincidence.

It sounds obvious, but to avoid the look-elsewhere effect, scientists should be careful to limit the questions they are trying to ask before they undertake the study in which they attempt to answer them. If well designed, the resulting study should prove sufficient to answer the questions they originally proposed and might also throw up other seemingly interesting relationships that suggest new questions to investigate. Because every data set will contain some patterns that are due to chance alone, it's not valid simply to draw a target around these patterns and conclude there is a relationship. If we want to find the true answers to the new questions suggested by these clusters, we need to collect more data to be able to undertake studies in which those new questions are on the list of things to investigate.

The Swedish study asked too many questions of its data set, with the ultimate outcome that none of the answers was trustworthy. To this day, much to the disappointment of Julie Larm and the Omaha parents for the prevention of cancer group, no credible link has been found between power-line electromagnetic fields and cancer – or any other detrimental health effect, for that matter. Although they ought to have known better, the Swedish scientists are by no means the only people to have fallen into the trap of probing too many different avenues.

The numbers game

At 6am on 21 March 1967, psychiatrist Dr John Barker answered the phone at the British Premonitions Bureau. On the other end of the line, an anxious-sounding man, Alan Hencher, rambled about a plane that would crash over mountains killing 123 or 124 victims. Less than a month later, a passenger aircraft came down during a thunderstorm over Cyprus. The next day the *Evening Standard*'s front-page headline read '124 Die in Airliner'.

Hencher's premonition seemed chillingly accurate. To have predicted the exact number of victims ahead of time seems so unlikely that we might consider the odds against the event sufficiently remote as to rule out the possibility of chance or coincidence. The only explanation, then, must be that Hencher genuinely could predict the future. However, when analysed more closely, there are a number of hidden factors surrounding this story which mean that the odds are perhaps not as unlikely as we might imagine.

Those who consider themselves clairvoyant commonly cite dreaming as a method by which premonitions are visited upon them. So let's try to estimate the probability of the seemingly unlikely event that someone has a dream that correctly predicts an aeroplane crash within the next month (the accident Hencher predicted occurred exactly thirty days after his premonition). Let's start by considering the 3.5 billion people in the world in 1967, each remembering roughly two dream themes per week[38] for the thirty days preceding an accident. This adds up to 30 billion dreams over that time period. Plane crashes are among the most commonly remembered dream themes,[39] but let's be conservative and suggest that they are the theme of 1 in every 1000 dreams. Even with this modest dream frequency, we'd still expect 30 million plane-crash dreams to have occurred in the month before a crash.

Hencher didn't just predict a plane crash, though; he seemed to predict the correct number of fatalities as well. This appears to demonstrate his precognisant power beyond reproach. But again, the sheer number of aviation-disaster dreams means that visions which correctly predict something as seemingly specific as the number of fatalities will be incredibly common. Let's assume that only one in ten plane-crash dreams is vivid enough for the dreamer to estimate a number of fatalities – that still leaves 3 million number-specific plane crash dreams to choose from. Given that the largest passenger plane at the time could carry no more than 260 passengers, any random stab at a number might have a one in 260 chance of being correct. Even reducing the number of candidate dreams by a factor of 260 suggests that seemingly precognisant dreams that also predict the correct number of deaths would still number over 11,000 in the month preceding the accident.

With his prediction of 123 or 124, Hencher gave himself a degree of leeway. In fact, as it transpired, two people later died of their injuries in hospital after being pulled from the wreckage, bringing the total number of fatalities to 126. That we are still impressed with the accuracy of Hencher's prediction, even though it was out by two, hints that, in our eagerness to be persuaded by a coincidence, we allow a degree of tolerance as to the exact details. A prediction of anywhere between 123 and 129 fatalities would, for many, still have been close enough to provide convincing evidence of Hencher's foresight, elevating his chances of success six-fold. This is an example of the *proximity principle*, which relies on close, but not perfect, matches being recognised as hits. The proximity principle is a favourite tool of conspiracy theorists and folklorists. It can dramatically improve the odds of a connection being made between otherwise unrelated events. Take, for example, the 'spooky' coincidences that urban legend suggests surround the assassinations of US presidents Abraham Lincoln and John F. Kennedy. Both the assassins, John Wilkes Booth and Lee Harvey Oswald, were Southerners, known by their three names and said to be born in the year '39. Even more amazing is the bizarre connection that Booth ran from the theatre where he shot Lincoln and was eventually caught in a warehouse, whereas Oswald ran from the warehouse from where he shot Kennedy and was apprehended in a theatre.

Taken at face value, these do seem like incredibly unusual coincidences, but when we dig a little deeper, we find that the proximity principle has been appealed to too liberally. It transpires that, as an actor, Booth was often billed as J. Wilkes Booth or just John Wilkes to distinguish him from the other actors in his family. Oswald was never known by all three of his names before the assassination of Kennedy, only afterwards. Because of his frequent use of false identities, including variations on his own name, Dallas police started using his full name for specificity. Although both were born in the South, after spending much of his life in the North, Booth regarded himself as a 'Northerner who understood the South'. The fact that both were born in one of two populous regions of the United States shouldn't inspire much awe. The coincidence of the birth year of the two assassins is rooted in misinformation. Oswald was born in 1939, but Booth, in reality, was born in 1838 not 1839. This inconvenience is typically swept under the rug during the retelling in the hope that no one will scrutinise the facts too closely. Finally, although Lincoln *was* assassinated in a theatre (one in which plays are staged) his assassin, Booth, was caught in a tobacco barn on a rural farm (not a warehouse). Oswald shot Kennedy from a book depository (which you could consider to be a sort of literary warehouse!) in the middle of Dallas and was later apprehended in a *movie* theatre. Sometimes, if you don't look too closely, near enough is good enough to be convincing.

Returning to Hencher's airline-disaster clairvoyance, with the six-fold leeway in fatalities afforded by the proximity principle, we might expect over 66,000 startlingly accurate aeroplane-prediction dreams across the world in the month leading up to a crash. Even if only 1 in every 10,000 people was in a position to report their dreams to the authorities, we'd still expect to hear of around six such predictions. Although Hencher may not have received his vision in a dream, including other potential means of experiencing visions would only serve to increase the number of such premonitions. All of a sudden, Hencher's ostensibly prescient prediction doesn't seem so unlikely anymore.

Accepting Hencher's prediction also requires us to gloss over the fact that Nicosia International Airport, situated in the middle of the Mesaoria plain at 220 metres' altitude, is hardly the mountainous region Hencher predicted for the crash. Hencher's success becomes even less impressive when we find out that it wasn't, by any means, the only prediction that he had recorded with the Premonitions Bureau – he made hundreds of calls in the first few years of its existence.

Hencher once called Barker at his home at 1am insisting that the psychiatrist check the gas supply at his residence because he was concerned about his safety. Barker's house wasn't connected to the gas mains. Ten days later, on 1 May 1967, Hencher phoned through another aeroplane-disaster premonition to Barker that he claimed would occur within the next three weeks – a good bet, given that, on average, there was a civilian plane crash every twenty days that year. As it transpired, that month – May 1967 – was the only one that year in which no fatal civilian plane crash was recorded anywhere in the world.

Hencher wasn't scared of going big with his predictions either. Two years later he made the following apocalyptic prediction:

Sometime before September 1969, a large lump of matter will be coming into space toward the Earth. Intense sunspot activity will reach an apex not known before. The combination of natural phenomena will cause floods, hurricanes, and severe earthquakes in various parts of the world. There will be approximately 500,000 dead.

Needless to say, you would have heard about it if this one had been correct.

The huge number of predictions that Hencher made over the years gave even this lone prognosticator a good chance of getting something correct in the long run. Allowing that we would be just as impressed if we heard the story of an accurate prediction from anyone on the planet, the probability of such 'premonitions' becomes almost certain. This is an example of *the law of truly large numbers* at work. It suggests that, for a given event, no matter how unlikely its one-off probability of occurrence, given enough opportunities, we should expect it to happen.

Although I made a fair attempt to estimate the probability of Hencher's plane-disaster prediction, I was forced to make a number of assumptions (although each time I tried to err on the side of making the prediction appear less likely by chance alone). For the vast majority of real-world coincidences, without going through such a mathematical argument, it is difficult to dissuade believers that the only agency behind the happenstance they perceive as causal is, in fact, chance. It is incredibly difficult to pin down, with any certainty, the mathematical likelihood of many such unusual occurrences. When trying to build up a numerical value for the likeliness or unlikeliness of an event, we are forced to make a series of basic assumptions, which, no matter how carefully justified, anyone sufficiently motivated can always chip away at, until the whole construction appears shaky, at least in the eyes of the believer.

There are, however, extremely unlikely events that occur all the time, which are more amenable to a solid mathematical treatment. We can precisely estimate the odds against such extraordinary episodes occurring and see exactly how the law of truly large numbers operates to make them far more likely than we might think at first.

It could be you (but it probably won't be)

It could be you was the advertising slogan of the UK's National Lottery when it first launched in 1994. Adverts encouraging people to splash out on a £1 ticket showed a giant finger-pointing hand descending from the sky to single out an otherwise ordinary individual for the life-changing jackpot. While the chances of winning big are pretty slim, as the old adage goes, your chances of winning increase dramatically (infinitely, even, if we're talking about relative probabilities) if you buy a ticket.

Mike McDermott was a firm believer in this reasoning. When the Portsmouth-based electrician checked his lottery numbers late in the evening of Saturday 5 October 2002, he couldn't believe what he was seeing. He had managed to match six of the seven numbers that came out of the drum, meaning that he was in line for a prize of over £120,000. For most people, this sort of windfall would be a life-changing sum of money, but by that point, Mike wasn't the same as most people.

In its earlier days, participants in the UK's National Lottery would pick six numbers between one and forty-nine (a so-called 6-from-49 choice) and could then watch live on national television as six 'main numbers' and a 'bonus ball' were drawn from the certified-random 'gravity pick' machine. Matching the six main numbers would win you a share of that draw's jackpot (or all of it, if you were the only winner), but matching any combination of five of the six main numbers and the bonus ball would also win you a substantial prize. The latter is what Mike had managed to achieve on that dark October evening.

The probability of winning the main prize is roughly 1 in 14 million. The first number out of the machine can be any one of the forty-nine balls. The next is chosen from the remaining forty-eight balls spinning around inside the drum, the next from forty-seven and so on, until the sixth, which is chosen from the remaining forty-four balls. The bonus ball is then picked randomly from the forty-three balls left in the drum. There are then $49 \times 48 \times 47 \times 46 \times 45 \times 44$ – or, if you work it out, over 10 billion – different ways in which six balls can be drawn from forty-nine in a specific order. The mathematical term for these different orderings is *permutations*. Many of these permutations will contain the same numbers as each other, just picked out of the drum in a different order. For example, the permutation 1, 2, 3, 4, 5, 6 is different from 6, 5, 4, 3, 2, 1, which is different again from 3, 4, 6, 1, 5, 2, and so on.

However, for most lotteries, the order in which the balls come out doesn't matter. When lined up in ascending numerical order, many of the potential draws will be repeats of each other. The mathematical term for a selection of numbers when the ordering doesn't matter is a *combination*. This term is used most commonly in a mathematical context when describing 'combination locks', like those used on bike chains or key safes. (Ironically, these should actually be called 'permutation locks' since the order of the numbers matters.)

To find the number of combinations from the number of permutations, we have to divide by the number of different ways that the chosen balls can be ordered. There are 720 ($6 \times 5 \times 4 \times 3 \times 2 \times 1$) ways in which to order six different balls (six possibilities for the first position, five for the next and so on, until there is only one choice for the last position). To find the actual number of distinct *combinations* of balls, we take the 10 billion ways of picking six balls from forty-nine and divide by the 720 orderings of six distinct numbers to arrive at roughly 14 million different possible combinations. Fourteen million to one, therefore, are the approximate odds of your chosen numbers coming up in the week's draw.

What Mike had achieved in October was actually slightly easier than winning the main jackpot. He'd also matched six numbers, but six out of the first seven rather than six out of the first six – five of the main numbers and the bonus ball. Since there are six different ways you can match five of the first six numbers (or equivalently, six ways that one of the first six numbers could be excluded) this meant that Mike's feat was six times more likely than winning the main draw.

Still, the odds against Mike's pick were longer than 2 million to 1. These seem like pretty staggering odds for a single individual to beat. However, given that between 20 and 65 million tickets were sold for each lottery draw in 2002, we shouldn't be surprised that someone beats these odds almost every week. In fact, assuming, at a conservative estimate, that 20 million independent, randomly chosen tickets were sold, the probability of someone *not* matching five numbers and the bonus ball is extremely small – on average, it should happen only once every 5331 draws. For the draw in which Mike won, sixteen other tickets also matched five numbers and the bonus ball.

The really surprising thing about Mike's win, however, was that it wasn't even the first time he'd done it. Just four months earlier, Mike had scooped almost £195,000 by matching five main balls and the bonus, playing with exactly the same choice of numbers. Now this really does seem unlikely. Mike himself said, 'We thought that winning twice with the same numbers would be simply impossible'.

Since all the draws are independent of each other, to find the probability of any one person matching five numbers and the bonus ball twice on any given pair of draws, we must multiply the probability of them doing it once (1 in 2,330,636) by itself to give a probability of less than 1 in 5.4 trillion. Professor Simon Cox from the University of Southampton told the *Daily Mail*: 'Thinking of something that is as unlikely as this is almost impossible. It is such an unlikely event that I can think of nothing else that is as unlikely – other than winning the lottery twice.'

Although it certainly is an unusual event, it's perhaps not quite as unlikely as the 'Five-trillion-to-one' headlines made it seem. For starters, as we've already suggested, given the huge numbers of people playing the lottery each week, we would expect tens of people to have a win like Mike's every single draw. Over hundreds of draws, this already puts the number of potential candidates for a second win in the thousands. Of course, the chance of these winners having a second success relies on them continuing to play the lottery. And why, you might ask, would they still be buying tickets after they'd had a big cash pay-out, since it seems extremely unlikely that they would win again? It's true that, never having won the lottery, the chances of winning twice are dramatically less than winning just the once. No matter how counterintuitive it may feel, however, once they've been successful in one draw, winners have in no way dented their chances of winning again and are just as likely to win as anyone else who plays (even playing the same numbers that they won with the first time).

The same reasoning about unlikely independent events underlies one of my favourite jokes. It goes like this: I picked up a hitchhiker last night. He seemed surprised about how readily I stopped and let him in. As we gathered speed, he asked me, 'Aren't you worried that I might be a serial killer?' 'No,' I answered. 'It's incredibly unlikely that there would be two serial killers in one car.' It's funny, of course, because you being a serial killer doesn't reduce the chances of the other person being a serial killer, even though it's correct to say the chances of two serial killers independently being in the same car are extremely low. In the same way, even though the chance of winning twice having never won are extremely small, winning the lottery once doesn't make you less likely to win again.

In fact, many first-time lottery winners have the perception that their chances of a second success are improved. They feel luckier having won once, which often induces them to play more frequently. As a case in point, even after completing the unlikely feat of winning *twice*, Mike told members of the press who had assembled to watch him spray champagne, 'People say that things always come in threes, so I will definitely be keeping my numbers. I now believe that anything is possible.' It's also the case that many lottery winners feel a sense of gratitude, and with their new-found liquidity can afford to buy more tickets than they might have done before their win, further increasing the chances of becoming a double winner.

So despite the fact that the odds of Mike's numbers coming up again are still over 2 million to 1 for a single draw, given that there are thousands of previous winners, many still playing and buying multiple tickets, it's easy to see that a second win for one of them in any of the multiple draws they have played since their first success really isn't so unlikely.

So what are the chances of someone winning the lottery once or even twice? Well, the law of truly large numbers says they are actually pretty good, but reason says that it probably won't be you.

Something perhaps even more unusual than Mike's double win happened in the Bulgarian national lottery in 2009. Each week, in front of an independent lottery committee, the Bulgarian lottery machine selected six numbers from a possible forty-two (a 6-from-42 lottery).

On 6 September, the numbers were 4, 15, 23, 24, 35 and 42. Four days later, exactly the same six numbers came up (although they were picked in a different order). The repeated draw made news around the world. The probability of a specific set of numbers appearing in two consecutive draws is less than 1 in 275 trillion. After the coincidence was noticed, a spokesperson for the lottery said, 'This has happened for the first time in the fifty-two-year history of the lottery. We are absolutely stunned to see such a freak coincidence.' Despite the claims of lottery officials that 'manipulation was impossible', the sheer improbability of the event led the Bulgarian sports minister, Svilen Neykov, to launch an investigation into the occurrence.

However, when we analyse the problem a little more closely, it turns out that the probability of a repeat draw isn't as unlikely as it may have sounded. The specific numbers that were picked the first time weren't really that important; the event only became notable when the same numbers came up again. Given that we'd be equally surprised if any set of six numbers was repeated, we really only need to consider the probability that the exact same set of numbers came up in the second draw, which is 1 in 5,245,786 for this 6-from-42 lottery – over 5 million times more likely than the 275-trillion-to-one odds that grabbed the newspaper headlines.

The same numbers coming up in consecutive draws is certainly surprising, but an exact repeat of *any* previous draw would likely still make the news. The Bulgarian lottery had been running for fifty-two years. Assuming two draws a week, this amounts to over 5400 independent sets of six numbers. Because we care about matching two different sets of six numbers, the really important thing is the number of pairs of draws. The number of ways to pair up two different draws increases like the square of the number of draws. With three draws there are three possible pairs. With 10 there are 45. With 100 there are 4950. With 5400 draws there are over 14.5 million pairs. The probability that none of these 14.5 million pairs is a match is just 6 per cent. In short, the overwhelming likelihood is that two sets of the same numbers would have been drawn at some point during the fifty-two-year history of the Bulgarian lottery.

Suspicions were aroused by the repeated draw when they probably shouldn't have been, given how likely it is to occur somewhere over a long enough period. No one won the Bulgarian jackpot the first time the numbers were drawn on 6 September, but people started asking questions when it emerged that a record eighteen people shared the jackpot from the second draw four days later. Something didn't feel right. As it happens, though, there's a perfectly innocent explanation for the multiple winners as well. Choosing the previous draw's winning combination is a surprisingly common strategy for many regular players, with those numbers typically being picked in the subsequent draw over 100 times more frequently than would otherwise be expected. The large number of winners from the second draw is exactly what we should expect.

Thinking about it rationally, if you were going to rig the lottery, you probably wouldn't tell seventeen other people about it, and you almost certainly wouldn't choose the same numbers that had come up in the previous draw. The fact that eighteen people played the numbers that had come up in the previous week, despite the fact that lottery results are independent of each other, probably says more about lottery players' superstitions and their lack of understanding of how to maximise their winnings (as we touched upon in the previous chapter and will investigate further in the next) than it does about potential interference. Each winner took home a disappointingly small £4600. When it concluded, the investigation into the repeated draw found no evidence of wrongdoing.

When you consider that there are hundreds of different lotteries worldwide, any of which I would be telling you about if they had experienced repeated draws, it becomes even less surprising that we have found one lottery in which the exact same numbers came out. As a case in point, on 16 October 2010 in the Israeli lottery, the same six numbers that had been picked just over three weeks earlier, on 21 September, came up in reverse order. Lottery officials initially revoked the result, fearing tampering or an error with the machine. Radio phone-ins were flooded with calls alleging that the draw was rigged, but again, after the draw was investigated, no evidence of any misconduct was discovered. When given enough opportunities, the law of truly large numbers suggests that even seemingly extraordinarily unlikely things can and do happen.

Combinations unlocked

The law of truly large numbers is often aided and abetted by combinations of interacting elements. When we were looking for the chances of two repeated draws in the Bulgarian lottery, we didn't just try to find out the chances of just one of the preceding thousands of draws matching the latest set of numbers. Instead, we looked at the probability of any of the thousands upon thousands of pairs of previous draws matching each other. When combinations ramp up the number of possibilities to the same size as the odds against the one-off version of the event happening, we should start to expect seemingly unlikely things to occur.

The maths we used to work out these combinations can have some surprising consequences. Imagine drawing cards from a standard deck of fifty-two at random and then putting them back. How many times do you think you would need to repeat this before the chances of having drawn the same card twice become more than 50 per cent? The answer is just nine. Do it nine more times and the probability is over 96 per cent.

Most people probably imagine that the four-digit PIN code that the bank sends them is unlikely to be shared by anyone they know – there are 10,000 permutations to choose from, after all. It turns out, though, that in a gathering of just 119 people, it's more likely than not that two people will have been given the same code. With 300 people, the probability increases to 99 per cent. Next time you're at a big enough gathering, try asking for the last four digits of people's phone numbers (for some reason, I've always found people reluctant to give their PINs to me) and see if you get a match between them.

The same reasoning suggests that in the 2065 draws of the old forty-nine-ball UK National Lottery (the UK has now switched to using fifty-nine balls), the probability of seeing a repeated draw was 14 per cent. This is small, but nowhere near impossible, although it didn't happen in all the twenty-one years that the lottery used forty-nine balls. It would have required just 4404 draws in total before the probability of seeing a repeat became more likely than not. Given that there are two draws a week in the UK, this equates to just over forty-two years of draws.

Probably the most famous use of this combinatorial maths is in finding out how many people you need to have gathered together before the probability that two of them share a birthday becomes more likely than not. The answer is surprisingly small, at just twenty-three.

With twenty-three people in the room, there are 253 possible pairs of people who might share a birthday. This large number of combinations means that even though the probability of any two *given* people sharing a birthday is small – just 1 in 365 – the probability that at least one *pair* of people (of the 253 pairs) in the room share a birthday is more than a half.

Combinations are frequently the driving force behind the huge numbers of possibilities that the law of truly large numbers relies on to generate seemingly unlikely events by chance. When there are enough possibilities for an event to occur, even if the chance of any one of them occurring seems low, together they can make even seemingly improbable events become overwhelmingly likely. When, for example, there are seventy people in the room, there are 2415 distinct pairs of birthdays to compare. The probability of finding a match among the almost two-and-a-half thousand pairs rises to above 99.9 per cent – near certainty.

The problem with our perception of these sorts of situations is that we are not intuitively very good at calculating the combinations involved and, subsequently, weighing their numbers against the small probability of an individual event occurring to see which one wins out. In the birthday problem, the number of pairs does not vary in proportion to the number of people in the room. Instead, it varies nonlinearly, increasing dramatically as the population of the room grows. As we will see from Chapter 6 onwards, we are not very good at thinking about nonlinear phenomena like this.

It is our underestimation of the enormous numbers of possibilities generated when elements combine that often makes the events (that the law of truly large numbers dictates will be quite likely to occur) so surprising when they transpire. As we have already seen, when something we intuitively considered to be unlikely happens by chance, we are prone to head off on a wild goose chase in an attempt to discern some underlying cause, where, in fact, there is none.

Illusory correlation

Illusory correlation is the name given to the phenomenon of perceiving a relationship when no such relationship exists. We are predisposed, for example, to notice seemingly meaningful clusters even in random data and then to infer an unwarranted connection. Spurious correlations and post-hoc rationalisation all prompt us to search for ways to avert looming crises or to replicate secret formulas for victory when, in fact, there was never a bullet to dodge or a reason behind the perceived triumph.

When assessing the likelihood of an unusual event, it's important to bear in mind our own deeply ingrained cognitive biases, so that we don't jump too quickly to erroneous conclusions. When an unusual event or coincidence befalls us, we must try to ask ourselves is this unlikely only from our individual perspective or from a population-level point of view. The answer will change the causal inferences we draw about the incident.

My friends and relatives have remonstrated with me in the past for ruining their fun – suggesting that their 'impossible' coincidences might not really be all that unlikely, thereby robbing them of their magic. For me, though, the magic lies in the recognition of these events as coincidences in the first place. Events that, at face value, appear to have impossibly long odds are actually occurring all around us all the time. We hear about the extraordinary stories of neighbours who find themselves sitting at adjacent tables in a café in a foreign city thousands of miles from home, neither having known that the other would be there; or of the adult who, browsing in a second-hand bookshop far from where they grew up, opens a copy of their favourite children's book only to find their own name scrawled inside in their own childish hand; or of the husband who, when looking through photos of his wife's childhood trip to Disney World, happens to spot his own father in the background pushing the buggy of a child who can only have been the husband himself.

Of course, we never hear of the neighbours who were seated on opposite sides of the café and so never caught sight of each other, the books you find inscribed with the name of a child from the next street who you never met or the husbands who don't look too hard at the background of their wife's childhood photos. The magic is not that seemingly extraordinarily unlikely events happen. Given enough opportunities, the mathematics dictates that they must. What makes these stories remarkable is that they are recognised at all. In the same way that understanding how rainbows form as sunlight is bent and reflected by raindrops only adds to their allure, in my opinion, the real buzz of the coincidence is in recognising it and then figuring out how on earth it could have happened.

More practically, it's important to be able to recognise coincidences for what they are – because improbable events can fool us into drawing the wrong conclusions about cause and effect, leading us to make generalisations that are not warranted, or to disbelieve facts that are staring us in the face. Staying rational in the face of coincidence, however, is easier said than done. Despite urging caution over seemingly unlikely coincidences, I find it hard to practise what I preach. Let me tell you about a coincidence which recently got me into trouble.

My wife's best friend from primary school, who she has remained good friends with for over thirty years, is called Tessa Hood. Tessa has a sister called Lucy. My daughter, who looks like a mini version of my wife, has a good friend in her primary school class who is also called Tessa. She, too, has a sister called Lucy. So far so interesting, I thought upon finding this out, but perhaps not that remarkable. One day, however, my daughter came home to tell me the wonderful coincidence she had spotted – that her friend Tessa and her sister Lucy also share the surname Hood. Two friends a generation apart, (one my wife's and the other my daughter's), both called Tessa Hood, each with a sister called Lucy.

Now, I knew for a fact (from a school WhatsApp group) that Tessa and Lucy's mum (my daughter's friends, not my wife's) had the surname Berry, so I assumed that my daughter had become confused when hearing about my wife's friend and her sister and conflated the two pieces of information. When I questioned her about this seeming discrepancy, she could tell, despite my attempts to be tactful, that I didn't completely believe her and she refused to talk to her doubting father about it anymore. We said nothing more about it, until, at a recent party, I met Tessa and Lucy's dad, who introduced himself as Ben Hood. I instantly felt bad about doubting my daughter.

This seemed like a remarkable coincidence. So remarkable, in fact, that I didn't believe it until I'd had it confirmed from the horse's mouth. In hindsight there were signs I should have seen, which would have made the coincidence seem less extraordinary, but which I either actively or passively chose to ignore. Tessa and Lucy are both relatively common names in the UK, so given how many families we know, both in and out of school, it's perhaps not that surprising that that name combination came up and that we got to know them. I was also happy to use the proximity principle liberally in describing the coincidence. My daughter's friend actually goes by the name Tess rather than Tessa, she also has a little brother whose existence I implicitly ignored when thinking about the sister coincidence. Lucy is also the older of the two Hood sisters for my daughter's friend, but the younger for my wife's. There were perhaps more differences between the two situations than I acknowledged when I started to scratch below the surface.

Nevertheless, I still think the coincidence is a good one, but perhaps not so good that I should have doubted my honest, intelligent and upstanding daughter. In hindsight, after I had apologised to her and been forgiven, the joy of the coincidence was in reconstructing the way she had put together the connection.

It just goes to illustrate that there are a huge number of reasons why apparently improbable events may not be as unlikely as they seem at first glance: the truly large numbers of combinations that can make a match almost inevitable; the seemingly impossible tricks where the cards were marked in advance; or the messages that we read in the static that were never really there at all.

The degree to which an event is remarkable is dictated by how surprising it appears. This, in turn, is directly related to our perception of the event's probability – the less likely it seems, the more surprising its occurrence appears. Being aware of the myriad mechanisms that lie behind the occurrence of seemingly improbable events can help us to better gauge their true probabilities and limit our unwarranted surprise without diminishing their wonder. If we recognise that the unexpected patterns we spot in the noisy background might just be part of the background noise itself, if we remain alert to the coincidences that might indicate a deeper unseen link between seemingly unconnected phenomena, but remember that they frequently do not, and if we can remind ourselves that even outrageously unlikely events happen all the time, then we will be better equipped to see beyond the lustre of illusory correlation when we meet these chance imposters face to face.

3

MASTERING UNCERTAINTY

In the previous chapter, we saw many examples of situations in which we were tricked by randomness into imagining underlying causation where there was really only background noise – randomness, rather than reason. If that chapter demonstrated how our cognitive deficiencies stop us from perceiving randomness correctly, this one will emphasise that our brains are limited in their capacity to generate randomness effectively for ourselves, which some suggest brings into question our capacity for free will and reduces our ability to make correct predictions.

Nevertheless, we will take what we have learned about randomness and use it to our advantage. We'll learn to spot situations which have the outside appearance of randomness, but which, in fact, are nothing of the sort. On the flip side of that same coin, we'll see how we can turn events that seem beset with uncertainty into apparent sure things. As we become more aware of our shortcomings, we will familiarise ourselves with strategies that allow us to make rational, future-facing choices in our encounters with uncertainty and even exploit randomness (or its absence) to help us make difficult decisions or maximise our winnings in games like the lottery.

How can you be certain?

In the last chapter, we discussed the surprisingly high probability of a shared birthday among a relatively small group of people. Since it featured in my last book, I've had the opportunity to experiment with the birthday problem with crowds of different sizes on many occasions. I often try to exploit the audience's misconceptions about the combinations in play by placing a small bet. Even with a relatively small number of people in the room, the large number of ways in which they can be paired up makes it overwhelmingly likely that two people will share a birthday. I always offer a good enough return on their stake to ensure an audience member won't be able to resist taking the other side of the bet – that there *won't* be two people who share a birthday – leaving me with the more likely option. I then go through, month by month, and ask people to call out the day of their birthday. With a large enough crowd, you can expect a pair in January. Sometimes, with smaller crowds, I've had to wait nervously until we reached December for a match, and once (annoyingly, while being filmed in front of a live audience at Google) I didn't find a pair at all and lost my money. On another occasion, having explained to the audience that with eighty people the probability of a match rises to 99.99 per cent and with 200 people it's above 99.9999 per cent, I was asked whether the probability ever actually reaches 100 per cent.

Approached from this angle, it seems like a reasonable question – will we just keep getting closer and closer to 100 per cent as the numbers increase? However, with a little more thought, the answer is not difficult to fathom. With 367 people in the room, you can be 100 per cent certain that there will be a match. There are only 366 different days that people can be born on (including the possibility of 29 February in a leap year). Even if the first 366 people you ask happen to have their birthdays on different days of the year, the 367th must have theirs on one of the days already taken. This is an example of what is known as the *pigeonhole principle*.

Imagine yourself working in a mail room with a hundred pigeon-holes. If, one morning, you have 101 letters to distribute, you can be certain that at least one pigeonhole will receive more than one letter. Try distributing the first hundred letters so that no pigeonhole has more than one letter. This means that each pigeonhole has exactly one letter in it. The final letter, then, will have to go into one of the pigeonholes that's already occupied. It's possible that if you distribute the letters a different way, several pigeonholes will have more than one letter and some will have none. It's also possible that all the letters go into one pigeonhole and none in any of the others. In any scenario you can think of, though, there will always be at least one pigeonhole with at least two letters. To generalise, if you have more 'objects' than 'categories' in which to classify those objects then some categories will have multiple objects assigned to them. Applying this *pigeonhole principle,* you can be certain that a given event will happen, even if, at first glance it seems hard to be sure.

Take for example the following question I ask the incoming maths students in my introductory freshers' quiz at the University of Bath: 'What is the probability that two people in London have the same number of hair follicles on their heads?' (I say 'hair follicles' instead of just 'hairs' to pre-empt the students who immediately point out that all bald people have the same number of hairs on their head – zero.) Typically, I get answers like 0.9999, or some other number of 9s repeating after the decimal point, because it seems extremely likely.

Rarely do the students reason their way to 1 – certainty. A reasonable estimate of the maximum number of hairs on a human head is around 200,000. Given that there are over 10 million people in London, the pigeonhole principle says that there aren't enough possibilities for them all to have different numbers of hair follicles on their heads. At least two must have the same number.

Another slightly less intuitive example comes from social networks. Imagine selecting a group of people on Facebook purely at random (it doesn't matter how large the group is, as long as there's more than one person in it – it could be as small as 10 or as large as 10,000). The pigeonhole principle says that at least two of the selected people will be friends with the same number of people in the group, but the reasoning is a bit more complicated. Bear with me. If there are N people in the group, then we can first recognise that there are N possible numbers of friend connections between people in the group, from 0 to N-1 (you can't yet be friends with yourself on Facebook). These are the N 'pigeonholes' into which all the N people in the group can be categorised. But for the pigeonhole principle to work, we need fewer pigeonholes than people. Fortunately, logic dictates that if someone is friends with everyone in the group, there can't be anyone who is friends with no one (and vice versa – if someone is friends with no one, then there can't be anyone who is friends with everyone), so there are, in fact, only N-1 'holes' in which to categorise the N people. The pigeonhole principle says that there must be at least one hole with at least two people in it. Equivalently, there must be at least two people who have the same number of friends in the group.

Because it can sometimes be counterintuitive and the reasoning is not always obvious, the pigeonhole principle is the basis of some excellent mathematical card tricks. Just as in the guess-your-number trick in Chapter 1, randomness seems to be at play when, in fact, the outcome is already determined by the mathemagician pulling the trick. The same idea – harnessing people's perceptions of chance – is at work in another classic mathematical hustle.

Perfect prediction, the stockbroker scam and the gullible gambler

On a stuffy Sunday evening in July 2014, I, like over a billion others around the globe, sat down to watch Argentina play Germany in the final of the men's football World Cup. As is often the case with these 'showpiece' events with so much on the line, it was a quiet and nervy affair with few clear-cut chances (especially in comparison to Germany's 7-1 demolition of hosts Brazil in the semi-final). At the end of an unspectacular regulation ninety minutes, the score was an equally uninspiring 0-0. Contrastingly, Mario Götze's wondergoal seven minutes from the end of the second period of extra time was a thing of beauty, devastating the Argentinians and winning the World Cup for Germany.

Although the match itself was nothing to write home about, I came across something far more striking a few days later. Twitter user @fifNdhs, alleging corruption in football's world governing body FIFA, offered 'proof' of match fixing by posting four single-sentence tweets on 12 July, with predictions about the following day's final. They read:

Tomorrows [*sic*] scoreline will be Germany win 1-0

Germany will win at [*sic*] ET [extra time]

There will be a goal in the second half of ET

Gotze will score

This seemed a truly remarkable set of predictions. To get one of them right would have been impressive, but by no means impossible. However, the combination of the four correct predictions in tandem from the same account seemed near miraculous. Unless, of course, the match *had* been rigged by FIFA and the events that would play out on the pitch were known in advance, as the poster led us to believe with a fifth and final tweet asserting:

Prove [*sic*] FIFA is corrupt

Although large-scale corruption within FIFA was eventually exposed just a year later, match fixing with that degree of detail in the most highly scrutinised game of football in the world was, in reality, beyond even their power. Instead, what @fifNdhs had actually done was a modern-day variation on a classic hustle known as *the perfect prediction, the stockbroker scam* or *the gullible gambler*.

In the traditional version of the hustle, the scammer writes to a large number of prospective victims or 'marks', giving the outcome of an upcoming sporting event or a prediction about the stock market. Usually, the predictions are about events with binary outcomes – a rise or a fall in a stock price or the winning team in a baseball match – and rely for their accuracy on a classic sports-prediction racket known as *double-siding*, whereby half of the marks receive one prediction and the other half the opposite, leading to a guarantee of correct predictions for half the punters. This structure neatly suggests starting with a number of potential marks that is a power of two – thirty-two, for example. After the first round, sixteen of the original thirty-two marks will have been given a correct prediction. Those sixteen are then followed up with another prediction about another stock price or sports match – eight with one result and eight with the other – while the sixteen who were given the wrong prediction in the first round are dropped. Again, pending the result of the event, the scammer then follows up the eight who were given another correct tip-off with predictions on a third event. Three correct predictions in a row might already be enough to allow the scammer to elicit a small fee for the next round of predictions from the four remaining marks. That fee can then be ramped up in subsequent rounds, as the marks are whittled to two and finally to one unsuspecting victim, who has now seen five correct predictions in a row – surely, they reason, too many for this to have happened by chance.

In 2008, sceptic and hoax exposer Derren Brown ran a real-life variant of the scam in his show *The System*. He successively whittled down 7776 (or 6^5) marks, eliminating five-sixths of the remaining candidates each round by predicting the outcome of a six-horse race. The effect of being given the correct prediction to five successive six-horse races, sometimes with extremely long odds, was enough to convince the final mark to pile her life savings on to Brown's prediction for the final race. Betting on all six horses in the final race himself and through some expert sleight of hand in switching her ticket, Brown was able to win that one for her as well.

I played a similar game at the end of March 2022 in anticipation of the publication of this book. I identified five events that were effectively two-horse races: England vs Australia in the final of the IPCC Women's Cricket World Cup; Oxford vs Cambridge in the boat race; Tyson Fury vs Dillian Whyte in a fight for Fury's WBC World heavyweight belt; Liverpool or Man City to win the English Premier League (the other teams were so far behind at this point that these were realistically the only two who could win); and either the Republicans or the Democrats to take a majority in the House of Representatives in November's mid-term elections.

I took £320 of my own money and bet £10 on all thirty-two (2^5) of the potential outcomes of the five two-horse races. The odds I received for the different bets varied from 2.25–1 (if the favourites Australia, Oxford, Fury, Man City and the Republicans all won), giving me a measly total return of £32.50, to roughly 920–1 (if the underdogs England, Cambridge, Whyte, Liverpool and the Democrats all won), giving me a return of £9216.80. I even made the effort to record thirty-two videos of me making the different predictions and showing the betting slips. I uploaded each one to my infrequently visited YouTube channel before the first event took place to date stamp the videos.

The first three events all went to the favourites – Australia beat England handily in the final of the cricket, Oxford won the boat race by over two lengths and Tyson Fury knocked out Dillian Whyte inside six rounds. As each new results came in I threw away betting slips and deleted videos, carefully pruning the evidence of my duplicity away.

The premier league title race was closer than the first three sporting fixtures. It went down to the final day of the season. With Manchester City two goals down at home to Aston Villa, it looked like Liverpool would take the title. But three goals in six minutes saw City turn the game on its head and clinch the title by a single point. Somehow having both sides of the bet covered didn't detract from the tension I felt watching the game as a City fan.

The race to control the House of Representatives was also closer than expected. The party of the sitting president almost always loses seats in the house – on average 27 seats. The fine 222-213 majority held by the Democrats since the 2020 election was expected to be easily wiped out by a swing of just five seats. Nevertheless, over a week after voting closed on 8 November and with only a few seats remaining to be called there was still no majority in the house. Eventually on 17 November 2022, the house was finally decided with the Republicans surpassing the unassailable 218 (of the 435) seats required to win an outright majority. Finally I had won my bet (not to mention the 31 other bets that I hadn't). Sadly with all the favourites having come home I only made £22.50 on my initial £10 stake.

Nevertheless my 'winning bet' stirred up some reaction when I posted about it on Twitter, with many people being impressed that I had correctly predicted the outcome of five unrelated events. Obviously I didn't mention the thirty-one incorrect predictions on that thread. Similarly to Nostradamus' 'predictions' that we met in Chapter 1, I relied on postdiction – the appearance of my correct bet only after the fact – to suggest that I have impressive prediction skills, when, in fact, I have nothing of the sort.

While my bets were just a bit of fun and everyone was left smiling at the conclusions of Brown's *The System*, in the real world, the ending to these sorts of tricks isn't usually so happy. The double-siding concept is reminiscent of the dishonest practices of some fund-management companies who create a range of 'start-up funds' – portfolios of stocks and shares designed, eventually, to attract external investors. For several years, the management companies 'incubate' these funds with small amounts of their own money, away from the public gaze. After the incubation period is over, the strongest-performing funds are aggressively marketed to investors and the weaker ones quietly culled to give the impression that the good performance was a result of skilfully chosen portfolios, rather than luck and covering a large number of bases. After being opened to the public, these previously high-performing funds generally fail to outperform otherwise equiv-alent counterparts.

This judicious pruning of poorly performing predictions is exactly how @fifNdhs achieved such a startlingly accurate set of World Cup final forecasts. The previous day, they had tweeted all manner of predictions, including 'Aguero will score', 'Kroos will score' and 'Argentina will win on penalties'. Their ruse might have been more successful had they made their account private prior to posting and deleting the erroneous tweets, instead of allowing them to be screen captured by another eagle-eyed Twitter user who was on hand to blow the whistle, but only after the fake predictions had garnered tens of thousands of likes and retweets.

*

These perfect-prediction hoaxes play on our prejudices, not least the authority bias we perceive to accompany a time-stamped social media post. Perhaps the most pertinent distortions we are subjected to in the above examples, though, are a pair of related but subtly different selection biases known as *reporting bias* and *survivorship bias*.

Reporting bias is characterised by the active suppression or selec-tive revealing of information. In particular, self-reporting bias is common in studies of lifestyle behaviours linked to disease. Surveys carried out to understand the epidemiology of sexually transmitted diseases, for example, are particularly prone to this sort of reporting bias. In one study on AIDS and sexual behaviour undertaken in France in 1992,[40] thousands of subjects were interviewed about their sexual practices over the telephone by more than a hundred different interviewers. The proportions of people engaging in risky or poten-tially illegal behaviours, like intravenous drug use, came as a surprise, being much lower than expected. One significant reason to treat the results with some scepticism is due to a form of reporting bias known as *social-desirability bias*, in which interviewees tend to respond in a manner which they think will be viewed favourably by others (in particular, their human interviewer). This bias is characterised by underreporting of what the subject perceives to be undesirable behaviours (e.g. having unprotected sex) and overreporting practices they think will be met with societal approval (e.g. condom usage). Paper-based questionnaires or automated telephone surveys remove human interviewers from the process, providing a greater sense of anonymity to participants. These autonomous survey techniques have been found to elicit significantly higher self-reporting in response to questions about 'sensitive' behaviours.

In an academic context, reporting bias is almost synonymous with *publication bias* – the phenomenon by which the outcome of a study influences the subsequent publication and visibility of that research. This is a huge problem when trying to assess the efficacy of treatments in a medical context. By surveying authors of papers on clinical trials about their unpublished work, researchers discovered that the results of trials with statistically significant results may be substantially more likely to be published than research in which a treatment has no discernible effect.[41] At first glance, it might seem reasonable to report only those results which appear to make a difference, but knowing which drugs don't work, as well as those that do, is of vital importance. When meta-analyses are carried out to assess the overall effectiveness of such treatments across multiple studies, the results can be distorted by publication bias,[42] leading medicine-regulatory agencies and doctors to believe that the drug is more efficacious than it really is. Litigation against pharmaceutical companies has uncovered systematic publication strategies designed to emphasise beneficial findings and diminish the appearance of the unfavourable impacts of their drugs. Attempts to reduce the extent of this malpractice have included prominent medical journals insisting that any research sponsored by pharmaceutical companies must be registered on a publicly available clinical-trials registry from its inception if the findings are to be reported in their publications. Despite these valiant attempts, there are always loopholes for sufficiently motivated researchers to squeeze through – delaying the reporting of negative results, publishing in languages that are less frequently used or in lower-circulation journals that are less likely to be read are all ways to cook the medical books.

If reporting bias is the deliberate suppression of information, then survivorship bias is perhaps its unintentional partner. It is often extremely hard to detect, since it depends on observers focusing too heavily, and usually unknowingly, on the items that make it through some, often unseen selection process, at the expense of those that do not. It's the reason I used to hear my grandparents saying, 'They don't make them like that anymore' in reference to some tool or other that's been passed down through the family. These surviving instruments don't necessarily prove that workmanship was significantly better in the past. Instead it's possible that many of the tools that weren't so well made didn't stand the test of time to be compared.

My favourite example of survivorship bias concerns the seemingly supernatural ability of cats to survive falls from high-rise buildings. Many believers in this urban legend cite a study which looked at cats taken to vets after falling from high-rise buildings in the late 1980s[43] (when apparently there was nothing better to study). Cats in the report fell from between three and thirty-two storeys. Remarkably, 90 per cent of the cats in the study survived. This result is paraphrased on www.pets.webmd and similarly on other pet websites as: 'There is a 90-per-cent survival rate for cats who are high-rise victims if they receive immediate and proper medical attention'. Adherents to this line, if they've thought a bit about it and know a little physics, may even venture to give an explanation for cats' amazing capacity to cheat death. The usual argument is to suggest that cats' ability to arch their backs allows them to – parachute-like – reduce their terminal velocity (the final speed an object reaches when the force of air resistance upwards is enough to counter the force of gravity pulling down-wards – not the other sort of 'terminal'), leading to more favourable landings. Intuitively, it sounds like an appealing theory. The other, perhaps more reasonable, explanation is that the study saw only those cats that were well enough to be taken to the vet. As one of my lecturers put it to me when explaining survivorship bias: 'You don't scrape many dead cats up off the pavement and take them to the vet'.

Psychic pet predictors

Survivorship bias is almost certainly also at play in another World Cup racket – the psychic pet predictors. Paul the Octopus made a series of astonishingly accurate predictions about Germany's results at the 2010 World Cup. Before each of Germany's matches, Paul's keepers at the Oberhausen Sea Life Centre would place two boxes containing food in his tank. One box bore Germany's flag and the other the flag of their opponent in the upcoming match. Whichever box Paul entered to seek his victuals was recorded as his prediction for the winners of the match (with no allowance made for the possibility of a draw). Paul's record at the 2010 World Cup was remarkable. He correctly 'predicted' all seven of Germany's games (including two losses), as well as the final between Spain and the Netherlands. Choosing the boxes at random, Paul's chances of correctly predicting the winners eight times in a row would have been 1 in 256 (or even longer odds, given that some of the group-stage matches could have ended in a draw). Putting your money where Paul's mouth was for the duration of the tournament would have no doubt earned you a tidy return on your initial stake.

And this was not the first time Paul had made successful football predictions. In the European Cup, held two years earlier, he had also 'predicted' four of Germany's six matches correctly using the same method. The probability of seeing Paul's overall record of twelve correct predictions from fourteen matches, assuming he lacked any psychic ability, would have been around 1 in 180. Those seem like pretty long odds and lend credence to the idea that this most intelligent of invertebrates must have had some sort of footballing sixth sense.

Sceptics have noted, however, that Paul's predictions were not made in a highly controlled environment. His box choices may have depended on the food in the box, the contrasting colours on the flags or Paul's position in the tank when the boxes were lowered in. Some have suggested that Paul was filmed multiple times with only the correct choice being presented to the public. However, given that Paul's predictions were made public before the outcomes of the matches were known, any suggested tampering would require a human handler being able to make the right predictions on his behalf which, although more probable, is still extremely unlikely. Although it's not clear how it would have made a difference to his predictions, some conspiracy theorists even suggested that the octopus that made predictions at the 2008 Euros died before the 2010 World Cup and was replaced by an octopus double, giving rise to a whole new 'Paul is dead' conspiracy.

Far more likely is that Paul's World Cup predictions were a result of chance and survivorship bias. Paul rose to prominence only after he correctly predicted Germany would beat England in the first knockout round – his fourth correct prediction – of the 2010 World Cup. Imagine that 200 animals started out making predictions (and what menagerie wouldn't put their most clairvoyant animal forwards given the potential prize of a huge surge in publicity and popularity like that experienced by the Oberhausen Sea Life Centre thanks to Paul's rise to fame?). The probability of at least one of these 200 animals making four correct predictions (allowing for draws in the group-stage matches), assuming that none of them has any special powers and they are just predicting at random, is 97.6 per cent. It's not surprising, then, that an animal like Paul rose to prominence. In fact, we would expect several such animals to have done so, and indeed they do at each World Cup. In this light, Paul's feat of predicting the remaining four matches, with a probability of 1/16, given that he survived the earlier rounds, seems far less remarkable. Indeed another 'psychic' pet, Mani the parakeet, also arrived at the final with a 100 per cent prediction record. Unfortunately for Mani, he picked the Netherlands, while Paul plumped for eventual winners, Spain, linking his name inexorably with the 2010 World Cup, while Mani faded into obscurity.

To illustrate how survivorship bias can combine with the law of truly large numbers that we met in the previous chapter, let me tell you about the 2018 World Cup. Mathematician and comedian Matt

Parker enlisted the help of over 1000 fans and pet owners around the world as he attempted to unearth some 'psychic' animals. In total, 133 pets started off predicting England's results. By the time they reached the quarter-finals, there were only two remaining with a 100 per cent record. Barry the Labrador correctly predicted that England would overcome Sweden in that round. And although he failed to foresee England's loss to Croatia in the semi-final, he did correctly predict their subsequent loss to Belgium in the third-place play-off match, leaving him with an overall record of six correct predictions from seven England matches – a record to rival that of Paul the Octopus.

It would sound remarkable if you didn't know how many other pets were making predictions from the start. Of course, I also deliberately failed to mention that Barry wasn't just predicting England matches. Of the fifty-nine fixtures Barry predicted over the course of the whole World Cup, he got just thirty results correct. My reporting bias, effectively lying by omission and only presenting a successful subset of Barry's predictions, made the canine clairvoyant look far more prescient than he really was.

Maximise your winnings

So it seems that betting on animals to predict football scores is not as good a strategy to make money as we might initially have hoped. As ever, picking the winner beforehand remains the difficult task. Nevertheless, many gamblers will tell you that it is the combined adrenaline and dopamine rush[44] that keeps them coming back, win or lose.[45]

Almost since its inception in 1994, this has been the attitude of many of the players of the UK's National Lottery. With the gamblers standing to lose 55 per cent of their stake on average each time they play,[46] there has to be something more than their expected winnings that keeps punters coming back to play, draw after draw.

On 14 January 1995 the UK's National Lottery was drawn for the ninth time. A week before, the eighth draw had ended with no one matching all six numbers. For just the second time in its history, the prize money rolled over to the following week, giving an enticing jackpot of £16,293,830. By the time the day of the draw came around, 70 million tickets had been sold – more than one for every person in the country. Large swathes of the nation were buzzing in anticipation of that evening's draw.

Pete Gallimore had played the lottery for the first eight weeks. Having won not so much as even the minimum prize of ten pounds so far, the car mechanic from Cinderford had decided to stop. However, upon reading about the size of the jackpot in the *Daily Express* that Saturday morning, he changed his mind. He walked back to the local corner shop where he had purchased his Saturday-morning paper and 'picked out some numbers', to hear him tell it, 'completely at random'.

That evening, as he watched the draw live on TV he was delighted, after seeing two of his numbers (23 and 38) plucked from the drum, when a third – number 17 – came out as well. He was guaranteed a prize, which would pay back all the stakes he'd sunk into the lottery so far. But then came a fourth, lucky number 7, and a fifth, number 32; and finally, a sixth number to match the remaining one on his ticket – 42. He'd won the jackpot. 'I couldn't believe it,' he recalls. 'I shouted to Janet to come in from the kitchen and got her to check the numbers with me. We both double-checked it on Ceefax before we dared to believe it was true. Sixteen million pounds; it was like a dream.'

In just a couple of days, however, the dream had diminished. Pete wasn't the only person to match all six numbers that evening. In stark contrast to no one winning the week before, Pete was confronted with the reality of having to split his sixteen-million-pound jackpot with others who'd matched all six numbers – 132 others, to be precise. Each of that evening's winners took home just £122,510 – still a large sum of money, but nothing compared to the prize each had thought they would receive when they first realised they had won.

'Gutted doesn't cut it. To think that you've won sixteen million pounds, only to be told it's less than 1 per cent of that is heartbreaking. For a few hours, we believed all our problems were over. We were planning for the good life. Don't get me wrong. It was great to win 100K, but it's just not what I was expecting.' The money enabled Pete to pay off his debts and to move to a bigger house, but he wasn't able to quit his job until five years ago, after his sixty-fifth birthday, when he officially reached retirement age.

Why had so many people picked the winning numbers for the draw in which Pete won just a 133rd share of the jackpot? The answer is precisely because we are so poor at being random. As we saw in the previous chapter, intuitively we think of random as being well spaced, so when we are asked to generate a random pattern ourselves, an even spread is often what we go for.

When Pete's winning numbers are marked on the National Lottery play slip, shown in the left-hand panel of Figure 3-1 below, several things quickly become apparent. Firstly, none of the numbers are adjacent to each other. The preconception that randomness somehow means spaced out seems to make consecutive numbers or clusters of numbers seem unlikely. Surely, we reason, dispersed sets of numbers are more likely to come up than six consecutive numbers. Indeed, dispersed picks *are* far more likely to be drawn than six consecutive numbers. It's tempting to conclude, then, that all choices with consecutive numbers should be avoided, but this reasoning fails. Well-spaced picks only occur more often because there are so many more well-spaced combinations than sets of six consecutive numbers. No single well-dispersed pick is anymore or less likely to come up than any other set of six numbers, consecutive or otherwise. As it turns out, mathematics dictates that almost exactly half of all draws in a six-from-forty-nine lottery should contain at least one pair of consecutive numbers.

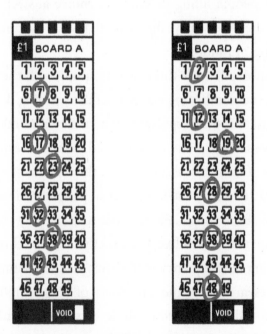

Figure 3-1: Winning numbers mapped out on the National Lottery play slip for the 133-way split jackpot draw on 14 January 1995 (left) and the fifty-seven-way split jackpot draw on 16 March 1996 (right). The selected numbers are well spaced (without being completely regular), non-consecutive and avoid the outer two columns.

The second thing to note from looking at the play slip is that each number is on a different row, but none of them is immediately above or below one another. This, perhaps, is because although we imagine randomness as being well distributed, completely regular spacing doesn't seem very random either, so small variations from a regular pattern help us to convince ourselves we are being random.

Thirdly, it's apparent that none of the numbers is drawn from the outer columns of the play slip. In fact, for this draw, all the numbers are drawn from just the second and third columns. This might be an effect of the middle bias we met in the previous chapter, which suggests that people filling in multiple-choice answer sheets – not dissimilar to the lottery play slip – tend to favour the middle columns when guessing 'at random'.

When combined, these three factors, along with the fact that the world's favourite number (7) was part of the winning draw, probably contributed to the over-representation of Pete's numbers in other players' choices. And if the most-shared jackpot wasn't enough to convince you of these biases, how about the second-most distributed top prize? On 16 March 1996, fifty-seven people won with the numbers 2, 12, 19, 28, 38 and 48. Plotting them on the play slip in the right panel of Figure 3-1 above illustrates exactly the same three principles.

By analysing the frequency with which jackpot-winning numbers are picked, researchers at the University of Southampton were able to demonstrate that all three of these biases are genuinely in play when people choose their lottery numbers.[47] They also confirmed the well-established proclivity for people to use birthdays as their picks – choosing the numbers between one and thirty-one with greater frequency than those above thirty-one. Picking according to these unwritten subconscious rules significantly reduces your expected winnings, precisely because you may have to share any jackpot you win with lots of other people. Choosing numbers purely at random is a better idea, but your expected returns are still only a meagre 45p for every pound you spend.

Although there's nothing you can do to increase your chances of winning the jackpot, there is a way to improve your chances of taking home as much of the prize as possible if you do win. Actively taking advantage of other people's randomness biases and choosing 'unpopular' sets of numbers can improve your expected returns to over 90 pence in the pound.[48] Still a loss, but a much smaller one than if you inexpertly try your hand at being random. Appreciating that we, as a species, are not very good at being random is the first step along the road to being able to recognise and avoid the fate of others who draw too readily on their misconception of randomness being evenly spread and unclustered.

Free will and free won't

Next time you're looking to kill some time with a couple of friends, you could try impressing them with the following deduction trick. Ask one of them to go away and flip a coin a hundred times, writing down whether it comes down heads or tails each time. Ask the other to write down another sequence of a hundred heads and tails, but one that they have generated for themselves without the help of a coin. Ask them to get together after they've completed their respective tasks and to make sure they know whose sequence is whose before mixing them up and handing them over to you. You can then amaze them by almost instantly telling them who generated which one.

The trick is not complicated. Have a quick look through each list. If there is a list of five or more heads or tails in a row, then you can be fairly confident that this is the genuinely random sequence. It's very unusual for a human to randomly suggest either five heads or five tails in a row. To our misguided minds, it just doesn't seem to be random enough. In fact, in a hundred flips of an unbiased coin, five heads or five tails has a 96 per cent chance of coming up.

This inability to be random has some deeper implications about the predictability of our own behaviour. If you google the phrase 'Aaronson Oracle', you will find a very primitive-looking website hosted by the University of Berkley as one of the top hits. The first line on the single-page website says, 'Press the 'f' and 'd' keys randomly. As randomly as you can. I'll try to predict which key you'll press next.' In effect, you are being asked to generate a series of heads and tails, just like the coinless participant in the trick I described above. I scoffed when I read this. How was this website going to be able to predict the key I would hit next when I was choosing completely at random? Although, in the previous paragraph, I suggested I could tell the human-generated list from the randomly generated list after the fact, I would never have attempted to predict what the participant making up the heads and tails would guess next. It would be impossible, wouldn't it?

After hitting the keys five times to give the algorithm some training data, the webpage started spitting out its prediction alongside the keystroke I just typed. After twenty-five keystrokes, a number appeared at the top to inform me of the percentage of my fs and ds that the computer had predicted correctly. After a little initial fluctuation, much to my dismay, the figure started to settle down around the high fifties. If I were genuinely hitting the keys at random, the best the computer should have been able to do in the long run is 50 per cent; the computer guesses should have been correct only half of the time. The fact that the algorithm predicted my keystroke more often than that meant that it was able to predict, to some extent, what I would do next. I figured this might be a fluke, so I refreshed the page and started again. After some initial wavering, the figure again settled down to just under 60 per cent. Finding this incredibly frustrating, I consciously tried very hard to be random, only to see the computer's ability to predict my choice improve. To some extent the computer 'knew' what I would do before I did it.

Still finding this automated clairvoyance hard to believe, I checked the code that was running under the hood to make sure the website wasn't just flashing up my choice as its prediction as soon as I pressed the button, with a few errors thrown in to make it look convincing. It wasn't. The algorithm it was using to make the prediction was quite straightforward. The code was keeping track of the keys I was pressing as it went on. Specifically, it was recording how often I had chosen each of the thirty-two possible sequences of five consecutive keys (f, f, f, f, f through to d, d, d, d, d). It then looked back at the previous four keys I'd hit and predicted the key I would press next, based on which of the two buttons would produce the more likely string of five according to the preferences it had learned from me.

This finding freaked me out a little. I started to wonder what this meant for my free will. Free will has been defined as the power to act on one's own conscious choice. If I was truly exerting my agency, shouldn't it be impossible to predict which button I was going to press next? Did I really have free choice when it came to which key to hit or merely the illusion of it?

More controlled scientific experiments have demonstrated the same phenomenon. Participants in one study were asked to write down lists of 300 digits (i.e. the numbers 1 to 9) in a random order. By considering the seven previous digits written down by the participants, the team of neuropsychologists conducting the study were able to predict the next one to occur with an average success rate of 27 per cent.[49] While 27 per cent doesn't sound hugely impressive, if the numbers were genuinely being generated completely at random, the expected success rate would only be 11 per cent. Our inability to act randomly may cast doubt over our capacity to make independent conscious decisions and might make us worry about the predictability of our actions by outsiders.

Who dares wins?

Rock paper scissors is often resorted to as a means to settle disputes, in much the same way that flipping a coin might be used to provide an unbiased randomised outcome. Indeed, when the president of the Maspro Denkoh electronics company decided to sell off his firm's $20-million art collection, he made Christie's and Sotheby's auction houses play rock paper scissors against each other for the honour (Christie's won by playing scissors against Sotheby's paper). But rock paper scissors is not the same as a random coin flip. It is a game played by human participants who, as we have just seen, are not very good at being random. As we will discover in Chapter 5, when we explore the intricacies of game theory, many games have strategies which lead to optimal outcomes if we can second guess what the other players are thinking. Rock paper scissors is no different.

Serious practitioners of rock paper scissors (as opposed to

squabbling siblings using the game to decide who gets to ride shotgun) understand that it is a game of psychology more than chance. Members of the World Rock Paper Scissors Society (yes, of course there is a society – why wouldn't there be?) are masters in the art of exploiting predictability. Anything but a completely random strategy can be analysed (just like the Aaronson Oracle analysed my keystrokes) and manipulated by a well-trained opponent to their advantage. Memorising a randomised list of papers, rocks and scissors to throw in a competitive game gives you the best chance of winning against a pro.

So if we really want to avoid the exploitation of our predictable behaviours, then ceding *some* control of our decision-making process to a 'randomiser' might be the answer. However, as in the extreme case of Luke Rhinehart, the protagonist of George Cockcroft's 1971 cult novel, *The Dice Man*, giving over *all* control might not be the best option.

Cockcroft describes how Rhinehart, a New York psychiatrist bored with his prosaic life, desirous of doing something daring but too afraid to deliberately choose to upset the status quo, decides to cede control of one simple, but momentous decision to the roll of a dice. If the dice lands on any of the numbers two to six he will carry on his evening as planned, clearing up after dinner and heading to join his wife in bed. However, if a one comes up, he will cross the hall to where Arlene – the woman he so frequently fantasises about – lives and attempt to sleep with her. Lo and behold, a one comes up. When Rhinehart returns to his flat after an evening in Arlene's bed, everything has changed for him. He begins to entrust every decision to the dice. The novel follows his randomised journey, as it unsurprisingly unravels and Rhinehart loses his reputation, his job and his family.

Inspired by the cult classic, in 1997, journalist Ben Marshall committed to an assignment that would see him follow Rhinehart down the rabbit hole into Diceworld. The job was not even Marshall's first foray into the world of dice-dictated decisions. At fifteen, becoming a voluntary dice-devotee, he had followed the dice to Brighton, where they dictated that he should lose his virginity to a prostitute, which he duly did. Later, as an adult, and now being paid for the experience, Marshall was even more committed to his randomised journalistic assignment. Over the course of two years, the dice encouraged Marshall to take heroin, cruise Santa Monica Boulevard for men and even introduce his girlfriend to the dice, resulting in her getting a job as a lap dancer on the Sunset Strip. Being a disciple of the dice, it seems, is not for the faint hearted.

The philosophy practised fictionally by Rhinehart and in reality by Marshall is a version of *flipism* – a popular literary trope under which decisions are made by consulting a coin or other randomising device. Villain Harvey Dent aka Harvey Two-Face from DC Comics' Batman franchise, for example, is almost entirely reliant on the flip of a coin to determine whether he will pursue acts of good or evil. While adhering to the strict tenets of extreme flipism certainly means decisions are taken, it doesn't always lead to the happiest of outcomes for the practitioner. There is evidence, however, that a small randomising nudge can be useful to break detrimental habits or to avoid getting stuck in a rut.

In February 2014, a strike on the London Underground Network caused significant disruption to hundreds of thousands of commuters, forcing them to find alternative routes to work. A team of economists from Oxford, Cambridge and the International Monetary

Fund analysed thousands of commuter journeys before, during and after the strike.[50] They found that up to 5 per cent of the people disrupted from their habitual commutes ended up adopting these alternative routes permanently after the strike had finished. Their results suggested that a significant proportion of commuters had failed, of their own volition, to tweak their journey to work to find the optimal commute. They'd chosen instead to settle for a passable trip which avoided the risk of the one or two extended commutes that might result from experimenting with their journeys. The economists predicted that the randomising factor of the strike, forcing some commuters to find better routes, would have a net economic benefit in the long term – the time saved by the 1 in 20 commuters breaking the habit outweighing time lost by all commuters during the strike.

Employing randomness to optimise a process is not a new idea. For hundreds of years, the Naskapi people of eastern Canada have been using a randomised strategy to help them hunt. Their direction-choosing ceremony involves burning the bones of previously caught caribou and using the random scorch marks which appear to determine the direction for the next hunt. Divesting the decision to an essentially random process circumvents the inevitable repetitive-ness of human-made decisions. This reduces both the likelihood of depleting the prey in a particular region of the forest and the probability of the hunted animals learning where humans like to hunt and deliberately avoiding those areas. To mathematicians, using randomness in this way, to avoid predictability, is known as a *mixed strategy*. We'll hear more about these strategies and others in Chapter 5 when we delve into the pragmatic and often counterintuitive world of game theory.

Analysis paralysis

Another way in which randomness can help us to make difficult decisions about the future is in the avoidance of *analysis paralysis*. If you're anything like me, then you might experience a mild form of this phenomenon when choosing what to order from an extensive menu. Should you go for the risotto or the burger, the steak or the pasta? I am so indecisive that the waiter often has to come back a few minutes after taking everyone else's order to finally hear my choice. All the choices seem good, but by trying to ensure I choose the absolute best option, I am running the risk of missing out altogether.

Restaurants are certainly not the only places in our modern world in which analysis paralysis rears its head. In almost all areas of life, from the everyday choices about the groceries we buy or the clothes we wear, to the big decisions concerning where we should live or whom we should date, the internet now delivers an unprecedented amount of choice. Even before the internet brought these decisions directly into our homes and the phones in the palms of our hands, choice had long been seen as the driving force of capitalism. The ability of consumers to choose between competing providers of products and services dictates which businesses thrive and which bite the dust – or so the long-held mantra goes. The competitive environment engendered by consumers' free choice supposedly drives innovation and efficiency, delivering a better overall consumer experience.

However, more recent theorists have suggested that increased choice can induce a range of anxieties in consumers[51] – from the fear of missing out (FOMO) on a better opportunity, to loss of presence in a chosen activity (thinking, 'Why am I doing this when I could have been doing something else?') and regret from choosing poorly. The raised expectations presented by a broad range of choices can lead some consumers to feel that no experience is truly satisfactory and others to experience analysis paralysis. That more options provide an inferior consumer experience and make potential customers less likely to complete a purchase is a hypothesis known as the *paradox of choice*.

In 2000, researchers from Columbia and Stanford Universities set out to explore exactly this hypothesis.[52] On two consecutive Sundays, they set up a tasting booth at a high-end grocery store in Menlo Park, California. On the first Sunday, the stall was stocked with twenty-four different flavours of jam, which customers were allowed to taste test. On the second Sunday, the number of samples was reduced to just six. The display with the larger range was successful at attracting 60 per cent of passers-by, while the smaller range attracted only 40 per cent. However, on average, the customers tried the same number of jams (just two) irrespective of the choice on offer. By far the most striking aspect of consumer behaviour seemingly revealed by the study came when the customers were followed up to find out how many of them had actually purchased a jar of jam. Just 3 per cent of those who'd been presented with two dozen jams ended up buying a jar, compared to an enormous 30 per cent of those who'd tasted from the six jams in the restricted range. The suggestion was that the excessive choice in the first day's products left consumers feeling ill informed and indecisive when making a purchasing choice.

The best is the enemy of the good

The bigger the decision, the more likely analysis paralysis is to strike. When we step back a little from the decision we are trying to make, it usually becomes clear that, although there may be one best option, there will also be several good options that we would be satisfied with – many houses, for example, where we would be content to make our homes and many people with whom we would be happy spending our lives.

I learned this 'good-enough' lesson for myself the first time I visited New York. I love climbing a high building to get a view of a new city, so I set off to do just that. But the queue that greeted me at the Empire State Building (at the time, the tallest building in New York) was so long as to put me off; so instead, I took the lift up to the top of the Rockefeller Center. The view was spectacular, and while not the tallest, 30 Rock still rises high enough to dwarf almost all other buildings in New York. Choosing an alternative that may not be the very best, but is at least good enough, has been christened *satisficing* – a portmanteau of *satisfying* and *suffice*. Psychotherapist Lori Gottlieb champions satisficing in her book *Marry him: The Case for Settling for Mr Good Enough*. When searching for a partner she advocates focusing on answering the question 'Am I happy?' rather than, 'Is this the best I can do?' As the Italian proverb that Voltaire recorded in his *Dictionnaire philosophique* goes: '*Il meglio è l'inimico del bene*' – 'the best is the enemy of the good'.

The idea that there is a perfect solution to a problem, particularly in subjective matters, is known as the *Nirvana fallacy*. In reality, there may be no solution that lives up to our idealised preconceptions – no perfect partner waiting out there for us, no dream house made real in bricks and mortar. Fortunately, randomness offers a simple way to overcome choice-induced analysis paralysis. When faced with a multitude of choices, many of which you would be happy to accept, flip a coin or let the dice decide for you. Sometimes, making a quick, good choice is better than making a slow, perfect one – or indeed being paralysed into complete indecision.

When struggling to choose between multiple options, having a decision seemingly made for you by an external randomising agent can help you to focus in on your true preference. Here, in contrast to Ben Marshall's dice experiments, you are not required to follow the decision of the randomiser to the letter. However, the externally suggested choice does serve to put you in the position of having to seriously contemplate accepting that option. This strategy can help you to envisage the consequences of what was, up until that point, a seemingly abstract decision. Experiments undertaken by a team of Swiss researchers have demonstrated that a randomly dictated decision prompt can help to deal with the information overload that often precipitates analysis paralysis.[53]

After reading some basic background information, three groups of participants were asked to make a preliminary decision about whether to fire or rehire a hypothetical store manager. After forming an initial opinion, two of the three groups were told that, because these decisions can be hard to make, they would be assisted by a single computer-generated coin flip. The coin flip would suggest whether participants should stick with their original decision (group one) or to renege (group two), but they were told that they could ignore the coin flip outcome if they wanted to. All three groups were then asked if they would like more information (an indicator of analysis paralysis) or whether they were happy to make their decision based on what they already knew. Once those who asked for more information had received it, all participants were asked for their final decision.

The participants who were subjected to a coin flip were three times more likely to be satisfied with their original decision and not to ask for more information than those who had not been exposed to the randomised suggested decision. The random influence of the coin had helped them to make up their minds without the need for more time-consuming research. Interestingly, requests for further information were lower when the coin suggested the opposite of the participant's original decision than when it confirmed their first thoughts. Being forced to contemplate the opposite standpoint made participants more certain of their original choice than when the coin flip simply reinforced their first decision.

It's comforting to know that when struggling to make a selection, we can allow a randomiser to make it for us. Even if we resolve to reject the coin's prescription, being forced to see both sides of the argument can often kick-start or accelerate our decision-making process. Diceman-like, divesting control of some of our everyday decisions to randomness can help us to find more efficient routes to work or just avoid the monotony of listening to the same songs by the same artists over and over again.

If we genuinely want to mix things up, though, it's important that we divest control to an external randomising agent, be that a die, a coin or a computer algorithm, because, as we have seen, we are not innately brilliant random number generators. Indeed, our inherent inability to be completely random has surprising implications for our abilities to be genuinely spontaneous. However, appreciating this fundamental flaw allows us to stay one step ahead of the crowd – for example, by keeping hold of more of our lottery winnings. A fundamental understanding of what uniform randomness really looks like can give us the upper hand in situations ranging from high-stakes business negotiations to (arguably) lower-stakes games that might decide who does the washing up.

On the flip side, as well as recognising, understanding and harnessing randomness for ourselves, we should be on the lookout for situations which appear to be more random than they actually are – the dead ends and failures that are hidden from us by the chance pruning of survivorship bias or the deliberate concealment of reporting bias. We must also be on the lookout for the calculated half-truths told to us by unscrupulous agents, inducing us ultimately to back the wrong horse. Like Sherlock Holmes, we must draw inferences not just from the information in plain sight but from the evidence that is missing – those curious incidents when the dog does not bark in the night-time. We should ask what it is we are not being shown. If we can begin to recognise that the glimpses of information we are being fed might not be representative of the whole picture, as well as being aware of the biases that might be present in the data we are shown, then we will provide ourselves with a degree of immunity against those who would seek to use our inherent struggles with randomness against us.

4

CHANGING YOUR MIND

Thus far, we have seen multiple examples of our innate inability to identify, comprehend and respond to randomness. In Chapter 2, we exposed the cognitive habits which tempt us into trying to pick out meaningful patterns in background noise. In many senses, this is completely the opposite of the problem with which we began the previous chapter, when we met situations which appeared on the surface to be governed by randomness, but which, in fact, were entirely predetermined. We also saw, in the last chapter, that we are not particularly good at generating randomness for ourselves – our capacity for genuine spontaneity being called into question as a result. However, we did begin to see how – if we were able to divest the creation of randomness to an external source like a coin or a dice – we could actively employ randomness to our benefit in some (but perhaps not all) decision-making processes.

Given the multiple potential hazards that the preceding chapters have highlighted, it's tempting to infer, then, that forecasting in the face of randomness is difficult or perhaps, in some circumstances, impossible. But this is far from the truth. Just because a behaviour is random, that doesn't mean that it is completely, or even largely, unpredictable. In fact, the probabilistic descriptions of many random processes contain numerous repeatable and reproducible characteristics. As we will discover throughout this chapter, mathematics can provide the X-ray specs that allow its practitioners to distinguish these telltale signals from the environment's inherent variability.

Perhaps even more importantly, as we will see towards the end of the chapter, mathematics can provide a framework for reasoning in the face of uncertainty. We can be very quick to lampoon people who change their minds with accusations of hypocrisy or indecisiveness. In some instances, the ridicule seems justified. When pompous politicians express convictions with 100 per cent certainty, leaving no room for doubt, and are subsequently proved wrong in short time, there is a degree of schadenfreude to be had in exposing their mistakes. But when people have taken genuine care to communicate the uncertainty associated with their original considered opinion, which they then update when confronted with new information, we should think better of mocking their course correction. Although rapid and complete U-turns are relatively rare, updating our views in light of new evidence is, after all, at the heart of science. As we attempt to navigate our way through the often uncertain waters of everyday life, we'll see that mathematics can provide us with the tools to help us steer our course – a compass that helps us determine how and when to change our minds.

Looking out for number one

To demonstrate the extent to which even seemingly random processes can harbour a degree of predictability, let me make a prediction about something personal to you of which I have no prior knowledge, and which may, at first sight, appear to be quite random. If you've got it handy, get out your address book. (In case you don't have yours to hand, I've reproduced house numbers from my own address book in Table 1.) If you've got a long address book, then perhaps consider just the first fifty entries. Go through the list and write down the first digit of the house number of each of your contacts. Now count up how many of these are either 1, 2 or 3. My prediction is that at least half of the house numbers in your address book will have a first digit that is either a 1, a 2 or a 3.

35	53	6	191	7	42	32	75	21	31	63	50	18
89	84	23	77	18	9	38	102	198	8	13	11	14
20	6	126	12	54	7	26	7	11	3	47	63	6
37	41	43	24	10	41	202	35	19	2	12	28	26

Table 1: Fifty-two house numbers from my address book

Counting the numbers of 1s, 2s and 3s in the first digits of the numbers in *my* address book shows that they account for thirty of the fifty-two entries – well over half. My guess is that the same is true for you. It might be a surprise that only three digits account for over half of all beginnings of the house numbers and that I can predict this with relative confidence. In fact, with only fifty entries from your address book, the probability that over half will begin with a 1, a 2 or a 3 is nearly 95 per cent. With a hundred, it approaches certainty, at 99.6 per cent.

This seems to run counter to our intuition that there should be the same number of addresses beginning with each digit, with no one digit being more or less likely than another. If this were the case, then each of the nine digits (remember no house numbers begin with a zero) should appear with a frequency of 0.11 or 11 per cent of the time. In other situations, this is manifestly the case. Imagine if there were a bias towards particular numbers that came up in the lottery draw. It would mean that each person's ticket had a different chance of winning. People would be able to game the system and some would stop playing altogether. The frequency with which the numbers are drawn must be evenly spread or, as a mathematician would put it, *uniform*.

For the first digits of street addresses, as with many other types of naturally occurring data sets, it turns out that the distribution is skewed far away from the uniform. The least common first digit is 9, at a frequency of just 4.6 per cent, and the most common is 1, at a seemingly remarkable 30.1 per cent of occurrences. The distribution which models such data types is known as *Benford's distribution* or *Benford's law*. You can see how well the first digits in my address book conform to Benford's law on the left-hand panel of Figure 4-1. The positions of the crosses match the heights of the bars pretty well. According to Benford's law, the digit 2 should turn up 17.6 per cent of the time. In my address book it appears with a frequency of 17.3 per cent. Four should appear 9.7 per cent of the time and I find it 9.6 per cent of the time. There are discrepancies, but these should dissipate with a larger data set.

To test this, I also got my hands on the addresses of over a million businesses in London. When I crunched the numbers (and believe me, there was some tedious crunching involved), I found the results shown in the right-hand panel of Figure 4-1. This time, the match between the predicted frequencies and the real frequencies was scarily good. The digit 1 is supposed to begin 30.1 per cent of the addresses and, in fact, it begins 30.8 percent; 3 should begin 12.5 per cent of the address numbers and, in fact, it accounts for 12.2 per cent of the numbers in the business database.

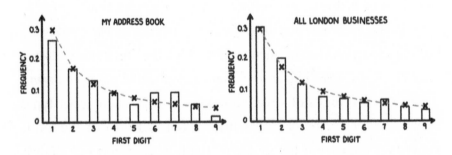

Figure 4-1: On the left, the frequency of the first digits of the fifty-two house numbers in my address book (bars) matches well with the frequency predicted by Benford's law (crosses joined by a dashed line). On the right, the frequency of the first digits of 1,008,925 London business addresses (bars) is an even better match to Benford's law (crosses joined by a dashed line).

Before you go out hustling for drinks in your local bar on the surprising strength of Benford's law, though, you need to know when and where it is OK to use it. We've already seen that if numbers are picked totally at random, as in the lottery, then Benford's law won't work. Perhaps mobile-phone numbers could be the basis of your hustle. The first three digits of phone numbers in the UK are prescribed, but perhaps the fourth holds true to Benford's law? Sadly, not. The digits 1 to 9 in UK mobile-phone numbers after the third digit are also uniformly distributed, meaning they are equally likely (a fact we used in Chapter 2 to calculate that with just 119 people it's more likely than not that two of them will share the last four digits of their mobile-phone number).

The distribution can't be too evenly spread. By the same token, if it is too constrained, then Benford's law won't hold either. If you look at data on the ages of Premier League footballers, or the height of adult males, or the IQs of ten-year-olds or the winning times in the women's Olympic 100 metres, you'll find the data are all clustered so tightly that they can only begin with a small number of different digits. There's no way Benford's law can hold. Indeed, many of these data sets are known to follow other well-known distributions. The bell-shaped normal curve, for example, approximately describes all sorts of phenomena, from the distribution of patients' blood pressure[54] to human male and female heights.[55]

However, mathematicians have been able to show that drawing data from many such different distributions, as the number of data points assimilated increases, the first digits of the data will display a very specific pattern – Benford's law.[56] So while many well-characterised phenomena, such as standardised test scores, lie under the control of one specific distribution, some more complex phenomena, which are made up of a random mix of many different factors controlled by many different distributions, conform to Benford's law. You could generate another Benford's-law distribution for yourself by going through unrelated articles in your favourite daily newspaper, finding those that mention one or more number(s), picking a number from each of them and calculating the frequencies of their first digits.

For Benford's law to hold we need the data to span several orders of magnitude. Thinking about the sizes of settlements in the UK, for example, reveals that there are many villages with fewer than a hundred people living in them and some cities with more than 10 million. Population sizes of UK settlements would be a good candidate for Benford's law to hold. Other similarly unconstrained data sets which conform to Benford's law include the numbers of pages in a publication, the heights of buildings and the lengths of rivers.

We've got your number

While the prediction I made earlier using Benford's law might make for a surprising party trick or win you a pint down the pub, it's unlikely to change the world anytime soon. But Benford's law has been applied in a whole range of real-world scenarios to detect artificially doctored data.

Defunct commodities company Enron was responsible for perhaps the most notorious case of accounting fraud in history. Systematic and institutionalised cover-ups allowed Enron to claim revenues of over $100 billion in 2000, a year before going bust. When financial data from Enron were analysed and compared to Benford's distribution, significant inconsistencies were found which might have alerted auditors to the fraud sooner.[57]

A group of four German economists scrutinised accounting statistics reported by all European Union member nations in the years leading up to the European Union sovereign debt crisis in 2010, which was precipitated, in no small part, by the Greek national debt crisis. When analysed, the Greek figures were found to have the highest divergence away from Benford's distribution, suggesting that they had been manipulated.[58]

Benford's law has been used to investigate irregularities in many different contexts, from presidential-election results in Iran[59] to fraudulent reporting in scientific research[60] and prosaically, but most commonly, in day-to-day accounts auditing.[61]

Wayne James Nelson was a low-ranking manager in the office of the Arizona State Treasurer. Over a ten-day period in October 1992, he wrote at least twenty-three bogus cheques to a third party, which he then diverted back to himself. Acting cautiously, he wrote just two cheques on the first day for the relatively modest amounts of $1,927.48 and $27,902.31. After waiting five days to ensure the coast was clear, he wrote four more and then, five days later, he threw caution to the wind and wrote another seventeen. In total, he cashed-in to the tune of nearly $2 million. But he made a big mistake. Although, to avoid suspicion, he chose the digits in each cheque to look random and had been careful not to duplicate any amounts or use giveaway round numbers, he'd been greedy. There was a limit of $100,000, above which internal accounting safeguards would mean his cheques would be examined in more detail. He'd been careful to stick below this limit in order to avoid unwanted scrutiny (writing later cheques for amounts which were close to but below the limit) – but he didn't know about Benford's law.

When his cheque amounts were analysed, Nelson was found to have stuck too closely to the $100,000 limit. Over 90 per cent of his cheques were for amounts above $70,000, compared to the roughly 15 per cent that would be expected from real cheque data that conforms to Benford's law. The probability of this happening naturally, at random, is roughly 1 in 1 quadrillion (10^{15}). Even after accounting for the large number of possible sets of cheques written each year, this low probability made the possibility that Nelson's behaviour was fraudulent seem extremely likely.

With no other option, Nelson admitted at his trial that he had faked the cheques – but he claimed that he had implemented his fraud to expose the lack of safeguards in his employer's new computer system. This flimsy excuse didn't stop him from being convicted and sentenced to five years in a state penitentiary.

Power is knowledge

One reason why Benford's law might crop up so frequently in everyday contexts is that many real-world data sets conform to a seemingly mysterious and more general law known as *Zipf's law*.

Zipf's law states that, for a large enough text, when the words are lined up in order of decreasing frequency, they exhibit a special pattern. Specifically, the second-most frequent word occurs roughly half as often as the most frequent. The third-most frequent word occurs approximately one third as often as the first, the fourth a quarter as often and so on.

I decided to test Zipf's law for myself, by analysing the word frequency of my previous book, *The Maths of Life and Death*. Lo and behold, when I completed the analysis, I found a startlingly good agreement with Zipf's law, which you can see in Figure 4-2. The most common word in the book is 'the', occurring 6691 times. Second comes 'of' with 3330 occurrences – almost exactly half the number of times that 'the' appears. Thirdly comes 'to' – at 2445 appearances, slightly over one third the frequency of 'the' – and so on. The word 'life' registered at number 145, on a par with 'mathematics' in position 146, both with 64 appearances, and well ahead of 'death' down in 230th place with 42 appearances.

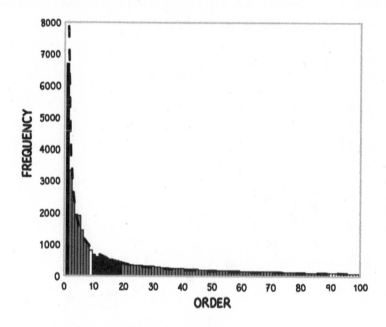

Figure 4-2: Number of appearances of the hundred most common words in The Maths of Life and Death *in order of frequency. Bars corresponding to words whose orderings begin with a 1 are highlighted in black and words whose orderings begin with a 9 are highlighted in white, while the rest of the bars are shaded grey. There are many more words whose orderings begin with a 1 than a 9. The black dashed line overlaid is the theoretical form predicted by Zipf's law, which matches the word-frequency data well.*

Figure 4-2 gives an insight into why data that conform to Zipf's law might also conform to Benford's law. In each order of magnitude (1–9, 10–99, 100–999, etc.), orderings which start with the digit 1 (black bars) are more common than those that start with any other digit, especially 9 (white bars). If the data follow Zipf's law and span enough orders of magnitude, we can be confident that Benford's law will also hold. When I analysed the leading digits of the orderings of the words in *The Maths of Life and Death*, I found that they conformed extremely well to Benford's law (Figure 4-3).

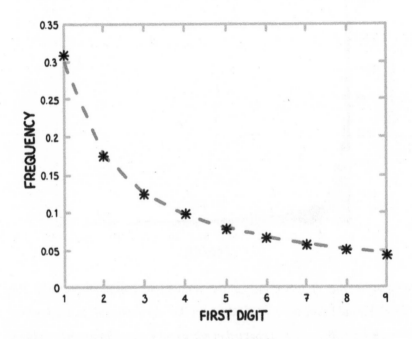

Figure 4-3: When the frequencies of the lead digits of the orderings of the words in The Maths of Life and Death *are calculated (black stars), they conform almost exactly to Benford's law (dashed grey line) – as will any data set which conforms to Zipf's law.*

Zipf's law for word frequency in a large text is universal. It doesn't just hold for English, but seemingly for many other languages – even the artificial language, Esperanto.[62] Fascinatingly, this almost magical relationship isn't simply limited to words in a text. It has also been found in extremely diverse scenarios, like the number of papers written by scientists,[63] the population size of settlements,[64] immune-related amino-acid sequence lengths[65] and even the diameters of craters on the moon.[66]

Zipf's law is a special case of a more general rule called a *power law*. Power laws, in this context, suggest that one variable (the strength of the pull of earth's gravity, for example) varies inversely with some other variable (the distance from the earth's centre) raised to some 'power'. For gravity, the shorter the distance from the centre of the earth, the stronger the pull, and the larger the distance, the weaker the pull. Zipf's power law for words in a large text is a special case for which the 'power' or exponent in the power law is one. This means that doubling one variable halves the other and that tripling the first decreases the second by one third, and so on.

For a general power law, however, this is not usually the case. The 'inverse square law' of gravitation, for example, follows a power law whose exponent (or power) is two. If you were to move twice as far away from the centre of the earth as where you are currently sitting, then the force you would experience at your new position would be four (2^2) times as weak as it is where you are now. If you move three times as far away, the force will be nine (3^2) times as weak and so on.

Power laws have been found to describe a wide range of naturally generated data sets, from the variation of species diversity with habitat area[67] to the frequency of the number of tornadoes per day in the United States[68] and even how the number of artists varies with the average price of their work.[69] Analysing data on wars from 1809 to 1949, Lewis Richardson found that the frequency of fatal conflicts varied with the number of people killed according to a power law[70] with exponent ½. Wars in which 1 million people died were found to be ten times less likely than wars in which 10,000 people died and a hundred times less likely than conflicts in which 100 people died. Perhaps one of the most important power laws ever discovered was published by Charles Richter and Beno Gutenberg in 1956[71] and purportedly has the power to predict earthquakes.

No great shakes

Cao Xianqing had been head of the Earthquake Office in Yingkou County, part of Liaoning province in Northeastern China, for just over four months. Having come late to education, first learning to read and write as a soldier in the People's Liberation Army, Cao was not your typical earthquake-prediction expert. What he excelled at, though, was organisation. After the region experienced a magnitude-5.2 quake in December of 1974, Cao set about organising a communication net-work, a transport team, a rescue team and earthquake offices in all the communes of Yingkou County. He started stockpiling winter clothing, bedding and food in preparation for a bigger quake.

After a number of false alarms in late December and early January, the region was hit by a series of small quakes in the first days of February 1975. Farmers in the region also began reporting changes in groundwater colour and levels, as well as strange animal behaviour, including 'frogs and snakes frozen on the roads' and 'rats that appeared drunk', as well as the more unimaginative 'neighing of horses and geese which frequently took flight'.

In the early hours of the morning of 4 February, a small magnitude-5.1 earthquake rumbled through Liaoning province causing some minor damage to buildings but nothing more serious. Cao snapped into action calling a meeting of the (Communist) Party Committee of Yingkou County at which he insisted there would be a large earthquake later in the day. Taking his enthusiastic remonstrations at face value, the committee ordered an immediate evacuation of all the communes in Yingkou County. Similar evacuation orders were relayed to other counties in the Liaoning province. Many people heeded the message and took refuge outside, away from buildings. Special showings of open-air movies were hastily organised to tempt reluctant citizens out of their homes on the cold winter's evening.

At 19.46, the big one – a magnitude-7.5 quake, with an epicentre on the border of Yingkou and neighbouring Haicheng County – struck. Bridges collapsed, pipelines ruptured and buildings toppled. Had no action been taken, it has been estimated that such a quake might have been expected to leave 150,000 of the region's 1-million-strong population dead in its wake. Yet, over the following days, when the dust had settled, just over 2000 bodies were pulled from the rubble. The rapidly implemented evacuation had saved tens of thousands of lives. This near-miraculous feat was quickly trumpeted by the Communist Party as evidence of China's ability to predict earthquakes.

The claim was very much at odds with the well-established consensus in the geological community that major earthquakes cannot be predicted. The United States Geological Survey (USGS), for example, even go so far as to state on their website: 'Neither the USGS nor any other scientists have ever predicted a major earthquake'. Had the Chinese cracked the earthquake-prediction problem and come up with a reliable method to accurately predict the location, timing and magnitude of earthquakes? Sadly, it seems the answer was no.

Some of the apparent omens of the 4 February earthquake could easily be explained as coincidence. The changes in the levels and colour of groundwater were directly attributable to a local irrigation programme. If you're going to raise the alarm every time horses neigh or geese take flight, then you are going to have a lot of false alarms. Indeed, the multiple false alarms in December and January (some of them instigated by Cao himself) are suggestive of the law of truly large numbers at work. If you make enough predictions about the timings of earthquakes, some of them will likely come true, but even more of them won't. Foreshocks might have been considered a reliable indicator of a potential earthquake, but as many big quakes are not preceded by foreshocks as ones that are. There are even more instances of lower-magnitude shocks which might be considered foreshocks, but are never followed up by a bigger quake.

Why was Cao Xianqing so convinced of the imminent threat that day that he risked his reputation instigating an evacuation of a region with a million inhabitants? Did he know something that everyone else didn't? As it transpired in later interviews with Cao, he did. Familiarising himself with local earthquake-education materials, he had come across a rule of thumb derived from a 200-year-old diary that stated, 'excessive autumn rain will surely be followed by a winter earthquake'. Having noted that it had rained more than usual throughout the autumn of 1974 and that 4 February was the last official day of winter in the Chinese calendar, Cao was convinced that the earthquake would strike before the day was out. But it was pure chance that it actually did.

Despite the apparent lack of science behind the fortuitous prediction, the Communist Party played up the role of their massed ranks of interested amateurs, like Cao, in monitoring precursory small-scale anomalies. This collective effort aligned nicely with Chairman Mao's Marxist ideology and allowed China to give the impression that they had cracked earthquake prediction. But that impression did not last very long.

On 28 July 1976, a magnitude-7.6 earthquake struck the city of Tangshan in Hebei province, directly bordering Liaoning. No one had predicted it. The earthquake struck at 3.42 in the morning, when most people were asleep in their beds. The majority of buildings in the city were so badly damaged that even if they remained standing, they were uninhabitable. Huge portions of the city's infrastructure were demolished instantly. The official death toll records a staggering 242,000 deaths, but seismologists studying the earthquake's impact have suggested this may be a significant underestimate.

It seems that individual earthquakes cannot be reliably predicted, but does this mean we can't say anything about the likelihood of a large earthquake occurring at a particular place over a particular period of time? Based on the earthquake frequencies in the two areas, I can make a prediction right now that my hometown of Manchester will experience fewer earthquakes of magnitude 4 or above than San Francisco over the next twelve months and be almost certain that I will be correct. This sort of future-facing projection is what the seismologists might call a forecast, rather than a prediction.

When the amount of energy an earthquake releases is plotted against the frequency with which those earthquakes occur, a distinctive power-law relationship emerges (see Figure 4-4). This is the celebrated Gutenberg–Richter law.[72] The data for earthquakes over the fifty-year period from 1970 to 2020 range from over 40,000 magnitude-4.5 earthquakes, releasing around 350 thousand million joules of energy each, to just two magnitude-9.1 earthquakes, releasing nearly 3 million million million joules of energy each. Because the two quantities (energy and frequency) vary so widely, the relationship is easier to see when plotted using logarithmic scales (as in the right-hand panel of Figure 4-4). When we do this, the data fall neatly on to the straight line predicted by the Gutenberg–Richter power law. The slope of the line, with value 0.7, is the exponent (or power) that characterises the underlying power law.

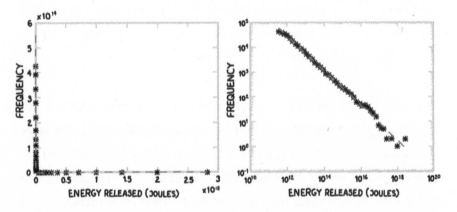

Figure 4-4 The frequencies of earthquakes around the world (1970–2020) vary, as a power law with how much energy they release. Because there are lot of relatively small earthquakes and few very large ones, the data in the left-hand plot tend to appear aligned with the two axes. The resulting relationship is easier to see when plotted using logarithmic scales on both axes. The characteristic straight-line relationship of a power law can easily be seen in the right-hand panel.

The Gutenberg–Richter relationship seems to indicate that earthquakes follow a very predictable pattern. Knowing how often smaller earthquakes occur in a particular region can allow us, therefore, to predict how often the larger and less frequent, but more deadly ones will occur. Although this doesn't allow us to specify the time, place and size of future earthquakes – what the USGS would refer to as a prediction – it does provide us with vital information that tells us whether the expected frequency of these events in an area makes it worthwhile expending time and money on earthquake preparedness.

With a 51 per cent forecasted probability of experiencing a magnitude-7 earthquake or higher over the next thirty years, for a city like San Francisco in a relatively wealthy country like the United States, it makes sense to invest significantly in earthquake preparedness. Even if the quake could be predicted precisely and all loss of life minimised, the economic cost of rebuilding the city's infrastructure after such a quake would be catastrophic in itself. Contrastingly, for relatively impoverished countries like the Philippines, a frequency of once every 450 years for magnitude-7 earthquakes in Manila, although higher than many other places around the world, might not justify the required expenditure to make the city quake-proof.

The historian Edward Gibbon wrote in his memoirs that the laws of probability are 'so true in general, so fallacious in particular'. Despite the fact that the Gutenberg–Richter power law appears to demonstrate that seemingly unpredictable earthquake occurrences can be spectacularly well behaved, it is a long way from being a crystal ball. It cannot foretell the precise date and time of the next big quake. Instead, it is limited to providing only the probability that a quake above a given size will occur in a given time period. Similarly, Richardson's power law describing the frequency of violent conflicts might tell you that the probability of there being no major war over the next twenty years is 90 per cent. This might make a bet against significant conflict in that period seem attractive, but it's worth remembering that you would lose your stake one time out of ten. This doesn't mean these forecasts are useless. Far from it. They allow us to prepare for a range of scenarios, allocating resources appropriate to the risk and likelihood of each. Exactly how we should trade off preparing for events with low probability but high potential for disaster against others with higher probabilities but lower danger is a question of *expected utility* or *expected payoff*, which we will investigate in more detail in Chapter 5.

When my information changes . . .

Interpreting the frequency of past events as probabilities is one of the best tools we have for managing uncertainty. This so-called *frequentist* view of probability allows us to make predictions about the future or inferences about the unknown current state of affairs. However, when new evidence comes along, we need a way of incorporating this into our world view and updating our beliefs. Fortunately, just such a tool for reasoning in the face of changing evidence has been around for almost 250 years. Now considered one of the most important tools across all of applied mathematics, *Bayes' theorem* (also known as Bayes' rule or sometimes just Bayes) has not always enjoyed this paramount status.

In the mid-1700s the amateur mathematician and Presbyterian

minister Thomas Bayes was operating in a climate in which probability theory was not well understood. Bayes wanted to know how to infer causes from effects and how to incorporate new evidence into his beliefs about a subject. At its heart, his theorem is a statement about conditional probability – the probability that a hypothesis is true given some piece of evidence. It might be the probability that a suspect is innocent (hypothesis) given a piece of forensic evidence, or it might be the probability (without looking at the team sheet) that Pelé was on the pitch (hypothesis) given that Brazil scored a goal (evidence). In real life, it is often easier to assess the transposed statements – the probability of seeing the evidence given that we assume an underlying hypothesis is true: the chances of seeing a particular piece of forensic evidence if a suspect is innocent or assessing the chances of Brazil having scored if Pelé was playing. Bayes developed his theorem as a bridge between these two sides of the conditional probability equation. He illustrated the sort of problem he felt his method could handle with a thought experiment.

Bayes first imagined his assistant playing at a billiard table while he himself sat in a room down the hall. His assistant would start the game by striking a ball so that it bounced from side to side across the width of the table. The assistant would then remove the ball, marking its position on the side of the table and challenge Bayes to discover the position of the marker. Clearly, without being able to see the marker Bayes would have no idea where it lay. To help him, the assistant would hit the ball across the width of the table so that its final position was equally likely to fall anywhere across the table's width. He would then call out to Bayes whether the ball had landed to the left or the right of the marker. The assistant would then repeat the process multiple times, each time telling Bayes whether the ball landed to the marker's left or right.

Given the position of the marker, finding the probability that the ball would stop to the left or right would have been easy. For a marker three-quarters of the way across the table, the ball would stop to the left of the marker with probability 0.75 and to the right with probability 0.25. But this was not what Bayes' assistant was asking him to do. Instead of asking him to find the probability of the ball being to the left or right of the given marker position, Bayes was being asked to find the more difficult side of the equation – the probability of the marker's position given that the ball fell to the left or right of it – a significantly harder problem, but one which Bayes' rule would allow him to tackle.

Bayes would start by assuming that the marker was equally likely to be found anywhere across the width of the table. His idea was that if the ball landed to the left of the marker, this should shift his expectation of where the marker was to the right and vice versa – the more times a ball stopped to the right of the marker, the more likely the marker was to be found towards the left-hand-side of the table. Each new piece of information allowed Bayes to restrict the probable region of the marker's location. Figure 4-5 shows, for increasing numbers of pieces of information, how the position of the marker becomes more and more certain as Bayes updates his ideas about its position.

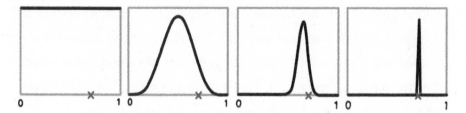

Figure 4-5: Bayes' estimates about the position of the marker (its true position denoted by the cross at 0.7) become increasingly accurate as he receives more information. The left-most panel indicates his prior belief that the marker could be anywhere across the table with equal probability. Ten pieces of information (second panel from the left) still give a distribution with significant uncertainty; 100 (second panel from the right) and 10,000 (right-most panel) pieces of information allow Bayes to become increasingly accurate in his estimate of the marker's position. Note that the scale of the vertical y-axis is different on each panel. We have not labelled the y-axis because we are primarily interested in the shape of the probability distribution – the narrower the distribution, the more confident we are about the position of the marker.

This was Bayes' idea in a nutshell: that he could update his initial belief with new data in order to come up with a new belief. In modern parlance, the *prior probability* (initial belief) is combined with the *likelihood* of observing the new data to give the *posterior probability* (new belief). As much as a mathematical statement, Bayes' theorem was a philosophical viewpoint: that we can never access perfect absolute truth, but the more evidence that accrues, the more tightly our beliefs are refined, eventually converging towards the truth.

*

Parts of Bayes' philosophy did not go down well in strict mathematical circles. In particular, the idea of a prior belief – that you should try to second-guess the answer to your question before you'd even performed an experiment or gathered any data – was unpopular. Prior beliefs seemed too far removed from the perceived objectivity of science, allowing too much room for flawed human judgment. That individuals with contrasting prior beliefs could reach different conclusions upon seeing the same data didn't seem to be consistent with the mathematical certainties to which the mathematical community had become accustomed.

Partly for this reason, Bayes' theorem never found consistent application during his lifetime. Even Bayes himself, it seems, failed to see its true importance, never bothering to publish it. His friend Richard Price, while cataloguing Bayes' notes after his death, discovered an unpublished manuscript containing the theorem. Appreciating its significance in a way which Bayes never had, Price spent some time editing 'An Essay towards solving a Problem in the Doctrine of Chances' before sharing the paper with the world.[73] Or so he thought. It turned out that almost no one actually read the manuscript in which the first formulation of Bayes' theorem was contained.

A decade or so later, the idea was rediscovered independently by the celebrated French mathematician Pierre Simon Laplace (who we will meet again in Chapter 9), who used it, among other things, to settle an age-old argument about sex ratios.[74] Starting with the assumption of each birth being equally likely to be a boy or a girl, he used Bayes' theorem to incorporate new information about sex ratios from France, then England, then Italy and, finally, from Russia. By updating his initial belief, he demonstrated that a sex ratio tipped slightly in favour of boys is 'a general law for the human race'. Surprisingly, even Laplace, who had used the formula with great success, eventually turned his back on the Bayesian way of thinking. The theorem fell in and out of use over the next 200 years, attracting both praise and derision from some of the greatest scientific minds.

Despite the continued scepticism and its unfashionable nature, there were many distinct successes during the period that Bayes' theorem spent in the hinterland. In the late eighteenth and early nineteenth centuries, artillery officers in the French and Russian armies employed it to help them hit their targets in the face of uncertain environmental conditions.[75] Alan Turing used it to help him crack Enigma,[76] significantly shortening the Second World War. During the Cold War, the US navy used it to search for a Russian submarine that had gone AWOL[77] (an event which inspired the Tom Clancy novel and subsequent film *The Hunt for Red October*). In the 1950s, scientists used Bayes to help demonstrate the link between smoking and lung cancer.[78]

The vital premise that all these Bayes adherents had come to accept was that it was OK to begin with a guess, to admit to not being certain of your initial hypothesis. All that was required in return was the practitioner's absolute dedication to updating their beliefs in the face of every piece of new evidence that came along. When applied correctly, Bayes' theorem would allow its users to learn from estimates and to update their beliefs using imperfect, patchy or even missing data. The Bayesian point of view does, however, require its users to accept that they are attempting to quantify measures of belief – to cast off the black and white of absolute certainty, and accept answers in shades of grey. Despite the paradigm shift required – thinking in terms of beliefs rather than absolutes – Bayesian reasoning didn't fit the subjective, anti-science label its detractors had pinned to it. In fact, Bayes absolutely typifies the essence of modern science – the ability to change one's mind in the face of new evidence. As the celebrated economist John Maynard Keynes once said, 'When my information changes, I alter my conclusions'.

Today, Bayes' theorem is at work behind the scenes, filtering out spam emails, from phishing attempts to pharmaceutical offers.[79] It underlies the algorithms that recommend films, songs and products to us online and is behind the deep-learning algorithms which are helping to provide more accurate diagnostic tools for our health services. Many of the theorem's more ardent disciples argue that Bayes' theorem is a philosophy by which to live. Although this is not my personal view, I think there are practical lessons we can benefit from if we learn to think in a Bayesian way – tools that help us to decide which of the multiple competing stories to believe, how confident to be in our assertions and, perhaps most importantly, when and how to change our minds. Bayes' rule has three main lessons for us to take away into everyday life, which I'll illustrate with a series of examples.

Lesson number one: new evidence isn't everything

You've moved into a new area and have been invited by a neighbour you've just met to their house for a party. At the party, the neighbour introduces you to a young man standing on his own in the corner. 'This is Paul,' your neighbour says. 'He's a m—' But just then, a crash comes from the kitchen and your host rushes off to clear up the mess, the end of their sentence lost to the noise. It sounded a bit like mechanic, but equally it could have been mathematician. Paul is so shy he will hardly look at you and mumbles when you ask him questions, so that you can barely hear his responses. In light of his behaviour, do you think it more likely that Paul is a mathematician or a mechanic?

Most people when posed this problem, even maths undergraduates learning about Bayes' rule, guess that Paul is a mathematician. Mathematicians have a reputation for being shy and socially awkward – a reputation not shared by mechanics. While, of course, there are many exceptions to the rule, stereotypes often reflect some degree of truth. As the old joke goes: Question: 'How do you know when you're talking to the extrovert at the Mathematics Department Christmas party?' Answer: 'They'll be the ones looking at your shoes, instead of theirs.'

Certainly, my lived experience as a mathematician suggests there is a higher predisposition towards social awkwardness and shyness than in some other professions, while most of the mechanics I've met are confident and unabashed when it comes to social interaction. If we guess that about half of all mathematicians and perhaps only 10 per cent of mechanics are socially awkward, this seems to provide a 5:1 ratio in favour of Paul being a mathematician.

However, there is another piece of information that feeds into the calculation, one that people tend to forget when answering the question. This is the relative numbers of mathematicians and mechanics out there in the UK – the prior probabilities of being a mathematician vs being a mechanic. There are around 5000 professional mathematicians working in maths departments in the UK (people who would likely describe their job as being a professional mathematician). This is dwarfed by the number of mechanics, at about 250,000. In assuming Paul is more likely to be a mathematician, we have made a classic mistake known as *the transposed conditional*. We've used our preconceptions of the probability of being shy, given that someone is a mathematician, instead of working out the probability of being a mathematician, given someone is shy.

Transposing conditionals occurs so frequently in courtrooms that it even has its own special legal designation: *the prosecutor's fallacy*. To commit the prosecutor's fallacy is to present the probability of seeing a piece of evidence if a suspect is innocent as the probability of the suspect's innocence having seen that piece of evidence. It's an understandable mistake to make because these two conditionals are related to each other through Bayes' rule. Their relationship depends on the prior probability of the suspect's innocence. If there are many plausible innocent ways in which the evidence could have been generated, then the two conditional probabilities can be very different numbers, giving very different perspectives on the probability of the suspect's guilt.

Figure 4-6 represents a scaled-down version of the Bayesian calculation we need to perform to understand whether Paul is more likely to be a mechanic or a mathematician. Half of all mathematicians are shy, but because the prior probability of being a mathematician is only about 2 per cent, those 2500 shy mathematicians account for only around 1 per cent of the total population of mechanics and mathematicians we are considering. Because the 250,000 mechanics account for almost all (over 98 per cent) of the combined population of mathematicians and mechanics, even 10 per cent of them – the 25,000-strong population of shy mechanics – account for almost 10 per cent of the total population. Shy mechanics (at 25,000) outnumber shy mathematicians (at 2500) by a factor of 10 to 1. This suggests that Paul is much more likely to be a mechanic than a mathematician.

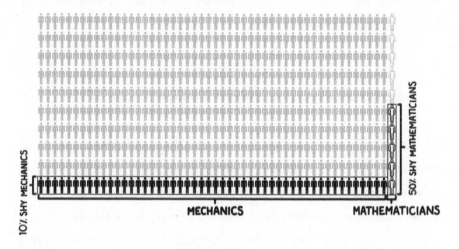

Figure 4-6: A scaled-down sample of 500 mechanics (black icons) and 10 mathematicians (outlined icons). To convert back to actual numbers each icon can be thought to represent 500 individuals. The outgoing (non-shy) mechanics and mathematicians have been greyed out as we are focusing only on the subpopulations which fit the evidence (i.e. are shy). Although 5 of the 10 (50%) mathematicians in the schematic are shy, mathematicians make up only 10 out of the 510-strong representative sample (less than 2 per cent). Even though a relatively lower proportion of mechanics are shy (50 out of 500 or 10 per cent), the mechanics' dominance of the population (500 of the 510 icons in the figure) means the number of shy mechanics (50) far outweighs the number of shy mathematicians (5).

Even if we lowered our estimate of social awkwardness in mechanics – let's say the percentages of shyness are only 2 per cent for the mechanics and still 50 per cent for the mathematicians – because the mechanics make up a far larger subsection of the group, the shy mechanics still outnumber the shy mathematicians. It's important to remember that new evidence shouldn't be the only thing which contributes to our beliefs. Instead, new information should be combined with our prior beliefs in order to update them.

Lesson number two: consider a different point of view

We already discussed earlier in this book how confirmation bias can lead us astray. The cognitive underpinnings of the phenomenon, however, are perhaps most neatly explained by thinking in terms of Bayes' theorem. Confirmation bias is, essentially, a failure to consider or assign sufficient weight to our prior beliefs about alternative hypotheses, or alternatively, an underestimation of the likelihood – the strength of evidence in favour – of these alternative hypotheses, or a combination of the two.

Imagine the situation in which you are trialling a new medicine to treat the chronic back pain you've been suffering from. After a week of taking the pills, you start to feel better. The obvious conclusion to draw is that the medicine has improved your back problems. But it's important to remember that there is at least one alternative hypothesis to consider. Perhaps your back pain fluctuates significantly from week to week anyway and, during the period over which you were taking the medicine, it's likely that your pain might have receded anyway. Perhaps less likely is the possibility that the improvement was caused by something else entirely – a different sleeping position or taking different forms of exercise, for example. We often fail to take this vital step back and ask, what if I were wrong? What are the alternative possibilities? What would I expect to see if they were correct? And how different is it from what I currently see? Unless we consider the other hypotheses and assign them realistic prior probabilities, then the contribution of the new evidence will always be disproportionately assigned to the obvious hypothesis we have in mind.

Alternatively, confirmation bias can arise when we are well aware of alternative hypotheses but fail to seek out, or assign appropriate weight to, evidence which contradicts our own preferred beliefs. This results in our overestimation of the likelihood of data supporting our favoured hypothesis and our underestimation of the likelihood of data supporting the alternatives. Twitter and other social media sites are classic examples of platforms on which many users exist inside an echo chamber. By being fed only those posts which reinforce their current views, their feeds shelter many Twitter users from alternative points of view. Users with what may start out as only mildly differing views have their opinions reinforced continually to the point of near certainty. This can result in increased polarisation and tribalism, both on the social media platform and back in the real world.

Much to my chagrin, the UK's 2016 referendum on membership of the European Union was a clear example in which I (and many others) completely failed to reason in a Bayesian manner. Almost everyone I had spoken to was voting to remain in the EU. Most of the people I interacted with on Twitter planned to vote in the same way. Despite polls having fluctuated between leave and remain in the lead-up to the vote, giving me a reasonably evenly split prior probability to work with, I assigned far too much weight to the anecdotal evidence I had gathered and failed to search hard enough for evidence supporting the alternative possibility. When, on 24 June 2016, I woke up to read the headlines that the UK had collectively voted to leave the EU by a margin of 52 to 48 per cent, I was genuinely surprised. I shouldn't have been.

We saw in the previous chapter how a random influence could be used to help us honestly confront our feelings about the consequences of alternative possibilities when facing a difficult decision. To avoid confirmation bias, we need a similar influence to help us to step back and actively search for, and equitably appraise, the evidence for all the different points of view, not just our favourite. If we are too enamoured with our own view, and consequently seek out and absorb only information which confirms it, ignoring any which disagrees with our preconceptions, we will end up drawing incorrect conclusions and making false predictions, even when presented with evidence which could have set us on the right track.

Lesson number three: change your opinion incrementally

Bayes' rule was never designed to be a tool that could only be applied once – to update a single prior with one new piece of evidence. Instead, as we saw in Bayes' original thought experiment at the billiard table, each new piece of information was used to update the most recent belief about the marker's position. The updated posterior belief was then used as the prior belief for the next round of the game and so on.

It is this ability to continually reuse Bayes' theorem to update our beliefs which skewers the objections held by the original probability purists that it is unscientific or unobjective to hold a prior belief about a subject. Very rarely will we have no information at all about a situation, no initial hypotheses which we can compare to each other to rate their likelihood. Science is a method in which we need to continually refine and update our ideas as new evidence comes in. Modern studies which might, for example, try to decide between the two hypotheses – whether anthropogenic climate change is a real phenomenon or not – should not be weighting these two hypotheses equally. The overwhelming weight of evidence so far suggests that it is real. That bias should be reflected in our prior probabilities.

We must be wary about overweighting our prior beliefs, too, though. The feeling of confidence in our convictions might make it tempting to ignore small pieces of information that don't change our view of the world significantly. The flip side of allowing ourselves to have prior beliefs as part of the Bayesian perspective is that we must commit to altering our opinion every time a new piece of relevant information appears, no matter how insignificant it seems. If lots of small pieces of evidence were to arrive, each slightly undermining the anthropogenic climate-change hypothesis, then Bayes would allow us to – indeed, dictate that we must – update our view incrementally.

Figure 4-7 reruns Bayes' thought experiment, but with an initial belief that the marker is over to the left-hand-side of the table and almost no probability assigned to it being in its genuine position (the cross at position 0.7). A single new piece of information makes an almost imperceptible difference to Bayes' prior belief. It would be tempting to disregard it. However, by the time 1000 pieces of new information have been accounted for, Bayes has a pretty good idea about where the ball is located. Enough small pieces of information can make a big difference. If Bayes had simply ignored the first piece of information since it didn't dramatically change his opinion, by the same argument he would have ignored all subsequent pieces of information and his prior belief would never have changed.

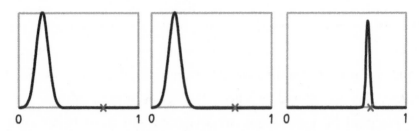

Figure 4-7: The evolution of Bayes' estimates about the position of the marker (denoted by the cross at 0.7) with an initial belief heavily weighted at an incorrect position. Bayes' prior belief is strongly that the ball lies towards the left-hand-side of the table (left panel). A single piece of new information does little to change that initial perception (middle panel). Incrementally, however, the impact of the evidence starts to accumulate in the prior belief. By the time 1000 pieces of new information have been accounted for, Bayes has a good estimate of the marker's position (right panel). Note, again, that the scale of the vertical y-axis is not consistent across the panels.

I will admit to believing that footballer Zlatan Ibrahimović was vastly overrated when he first came to play in England in 2016. Despite already being one of football's most decorated players and having played for some of Europe's biggest clubs, following Zlatan's achievements only intermittently had allowed me to write off each new domestic trophy, forgetting just how many he had won before. My prior belief, that he was not all he made himself out to be, was strong enough that even his spectacular bicycle kick against England in 2012 did little to persuade me of his quality. I was not behaving as a Bayesian.

However, when Ibrahimović moved to England to play for Manchester United – my own club, Manchester City's, biggest rivals – I couldn't help but begin to update my opinion incrementally. Watching each subsequent goal fly in on *Match of the Day*, my opinion of Ibrahimović slowly began to improve. Watching a player who was clearly in the twilight of his career, scoring eighteen goals in just three months and becoming the oldest player ever to score more than fifteen in an English Premier League season, left me with a significantly higher opinion of Ibrahimović than I began with (although still nowhere near as high as his opinion of himself).

Incrementally, small pieces of evidence can start to change the way we feel about an issue. We might have ignored them individually, because they appeared to make little difference, but slowly, the small shifts of position caused by the gradual plate tectonics of data acquisition can accumulate, until we find ourselves raised up to the top of an evidence mountain.

It's not always an easy thing to do, to change our opinions in light of new evidence. It feels uncomfortable to admit we were wrong and almost cowardly to renege on the beliefs we previously held on to so strongly. In fact, it requires great courage to hold and to espouse a contradictory view to one you have previously embraced. If, in political circles, there were more tolerance of decision makers changing their minds in the light of new evidence, rather than being harangued and derided for 'flip-flopping' and 'U-turning', then perhaps more of our policy makers would adopt a Bayesian approach to providing evidence-based policy, rather than sticking to their original guns, which now point squarely towards their feet, rather than to the target.

Of the three rules of thumb that Bayes' theorem suggests – that new evidence isn't everything, that we should consider different viewpoints and that we should change our opinions incrementally – we are perhaps most instinctively able to employ the last: to update our opinion based on our everyday encounters.

Unfortunately, this natural ability to learn from experience in a Bayesian way can sometimes be limiting to our aspirations. Experience can reinforce our opinions of our own self-worth – what we should be paid, how we deserve to be treated by our bosses, where we see our place in the world – until we come to believe that this is just the way the world is. Just like Bayes' increasing confidence about the position of the marker in his billiard-ball experiment, the more reinforcing evidence that accrues, the further entrenched our status quo becomes. The situation becomes almost a *self-fulfilling prophecy* (more of which in Chapter 7).

But what happens when the marker changes position? Because we have so much reinforcing experiential information, unless we can somehow reset our prior beliefs, it will take a long time and a lot of evidence before we can adjust our posterior beliefs – our view of our place in the world.

This is a particular problem in my own subject, as well as in other industries which struggle for representation of women and minority groups. If, as an aspiring black maths student, you have never come across a black mathematician, what do you think your prior belief about the probability of becoming a mathematician yourself would be? Probably pretty close to zero. With a low prior expectation, it becomes difficult to convince even the most promising students that they belong in a particular industry or in a given position. This is just part of the reason why improving representation is so important if we hope to increase diversity in industries traditionally dominated by particular sections of society.

Dealing with uncertainty

If we let it, Bayes can be a powerful tool for updating our preconceptions in the light of new data. What Bayes doesn't do for us, however, is suggest how we should pick those prior beliefs in the first place. There will always be some people who hold their convictions with 100 per cent certainty – think of religious fundamentalists, anti-vaxxers or climate-change deniers. What Bayes' theorem tells us in these situations is that there is not a single piece of evidence, no matter how strong, that will ever shift these hardliners from their convictions. There is very little point in asking someone whose prior belief is 100 per cent certain that vaccines work to debate with someone who holds that there is a 0 per cent probability. Neither stands any chance of changing the other's mind. As is plain from studying the many sectarian and religious conflicts throughout the centuries, when unstoppable forces meet immovable objects the results are seldom pretty.

At a societal level, it's important to understand that, in many contexts, random does not mean inherently unpredictable. Many real-world phenomena that at first appear to be governed by unadulterated randomness, including earthquake frequencies, income distributions or conflict sizes, actually obey highly replicable laws at an aggregate level. We can exploit these relationships to make forecasts which allow us to plan for the future, even if we cannot predict the exact timings and locations of the individual events themselves. Appreciating that 'random', does not always mean completely without predictability can allow us first to extract these characteristic signatures of the noise, to detect when someone is faking them or to use them to make plans for the future.

In the next chapter, we will see examples of how randomness can be exploited when developing strategies to employ against opponents in competitive scenarios. Thinking strategically through the lens of game theory will allow us to gain insights into what our opponents are thinking. Formalising the rules of engagement for ourselves in any given situation will help us to decide the most appropriate way to proceed, or even to determine whether we can alter the rules of the game to improve the outcome for all concerned. We will explore how we can employ randomness in so-called 'mixed strategies' that prevent our actions from becoming predictable and help to keep our adversaries on their toes.

Randomness – experienced passively in the world we are constantly trying to interpret or exploited actively in the decisions we make – is a fact of life. Rarely, if ever, is anything we do completely within our control. External forces, be they with known or unknown causes, often intervene capriciously, landing us in situations we could not reasonably have foreseen. Mastering the art of reasoning in the face of uncertainty while accepting that we will not always make the right choices, generate the correct predictions or hold the most well-grounded opinions, is one way to cope. Ultimately, we will all be happier once we learn to accept, if not always expect, the unexpected.

5

PLAYING THE GAME

Since 1950, the Egyptian-controlled Straits of Tiran had been closed to Israeli traffic. The narrow shipping lane, separating the Sinai and Arabian Peninsulas, provided Israel's only access to the Red Sea proper, and beyond that the Arabian Sea and the Indian Ocean. With a primary objective of forcing the reopening of the Straits, in October 1956, Israel invaded Egypt, kick-starting the Second Arab–Israeli war. Israel's covert allies, Britain and France, subsequently followed suit in an attempt to wrest control of the recently nationalised Suez Canal back from Egypt. The disastrous and short-lived 'Suez Crisis' that followed brought extreme international pressure to bear on all three invading allies. Britain and France were forced to withdraw embarrassed. Israel also pulled out, but with the key concession that the Straits of Tiran would be reopened to them indefinitely. Although the United Nations deployed an 'Emergency Force' along the border between Egypt and Israel, neither side took steps to ease tensions.

Fast forward eleven years to 1967. A number of incidents between Israel and its surrounding Arab neighbours reawakened the tensions in the region that had been slumbering noisily for the previous decade.

A military skirmish with Syria, ignited by perceived encroachments from Israel into the demilitarised zone between the two countries, led Israel to make then unsubstantiated threats to invade Syria and overthrow its government. Against this charged backdrop, a Soviet intelligence report was fed to both the Syrian and Egyptian governments. The report suggested that Israeli troops were amassing along the Syrian border ready for an imminent invasion. In fact, the information contained in the report was suspect, but Egypt's President Nasser couldn't afford to risk inaction. His popularity had already suffered significantly after having failed to react to Israel's recent attacks on Syria and Jordan. To exert his authority, in May 1967, Nasser remilitarised the Sinai Peninsula bordering Israel and expelled the UN peacekeepers from the region.

Recognising the potential threat to Israel's strategic position (90 per cent of the country's oil exports passed through the Straits of Tiran), Israel's Prime Minister Eshkol reiterated his country's post-Suez stance – that the closure of the Straits to its ships would amount to a threat of war. Despite knowing that it would make conflict almost inevitable, Nasser went ahead and blockaded the Straits, falsely claiming to have mined the watercourse.

The prospect of war was now tangible, with both sides anticipating the potential advantage in being the first to attack. Contrary to Nasser's favoured position of forcing Israel to be the aggressor and begin the conflict, his field marshal, Amer, told one of his generals in late May: 'This time we will be the ones to start the war'. An attack which would see the Egyptian Air Force bomb Israeli ports, cities and airfields was planned for the morning of 27 May 1967. Just hours before the planned strike, however, Nasser received a message from the Soviet Premier, Kosygin, suggesting that the Kremlin would not support Egypt if it began the war. Messages to delay the attack arrived only just in time, reaching pilots who were already in their planes awaiting final confirmation of the attack.

In the end, when it came, nine days later, on 5 June, the flame which lit the touchpaper of war came from Israel. The Israeli military struck with devastating speed and finality, catching the Egyptian forces unawares, as they launched a ground offensive into the Egyptian-occupied Gaza Strip and the Sinai Peninsula. At the same time, a comprehensive bombing raid wiped out nearly all of the Egyptian Air Force. Despite the Jordanian and Syrian militaries entering the fray against Israel on different fronts, in honour of their defence treaties with Egypt, Israel successfully defended the new attacks and conquered the whole of the Sinai Peninsula in just a few days. During the hostilities, Israel lost fewer than 1000 troops, while the combined losses of the Arab nations numbered over twenty times that figure. Having had their forces severely disabled by Israeli attacks, Egypt, Syria and Jordan signed ceasefire agreements just six days after Israel's initiation of hostilities. The 'first-strike advantage' carried by Israel into the 'Six-Day War' had proved to be decisive.

The rules of the game

In hindsight, it seems that the war was inevitable from the moment Nasser felt forced to blockade the Straits of Tiran. But was it? Could there have been a peaceful resolution through negotiation? Would it have been possible to reach a settlement to everyone's advantage when compared to the costs each side might have expected to have incurred through going to war? To find out, we will take a journey into a popular, but relatively recent, future-facing branch of mathematics: game theory.

It seems perverse to think of understanding something as consequential as international conflict by analysing trivial games when the outcomes of such hostilities could hardly be more serious. But game theory embodies a principle fundamental to mathematical modelling – that of simplification. When we strip away as much of the extraneous detail of a situation as possible, we are better able to focus on what's left behind – the bare bones of the problem. Many situations involving conflict can be reduced to relatively simple sets of rules, with the same underlying dynamics as the games we might play around the kitchen table. Time and time again, game theory has proved its worth as a field of predictive science. At the very heart of the approach is the assumption that the parties competing with one another are rational and always act in their own self-interest. And believe it or not, these two intertwined assumptions almost always hold.

It's reasonable to question how useful game theory really is by coming up with scenarios which, on the surface, seem to negate its fundamental assumptions. We can all think of situations in which the protagonists don't appear to be acting in their own best interests. How could we use game theory to predict and prevent a suicide bombing, for example, when the actions of the bomber are so clearly irrational? Who in their right mind would kill themselves in order to deprive other people of their lives? Once dead, what benefit could they possibly derive?

To answer these questions, we must be careful not to become too enamoured with our own world view. Because most of us abhor murder and place a high value on our own lives and livelihoods, it is easy for us to mischaracterise the actions of suicide bombers as irrational. But to really exploit the benefits that game theory can offer, we must put ourselves in other people's shoes and attempt to see the world from their perspective. An appreciation of what motivates the player in any given game is fundamental to applying game theory in the real world.

The key to using game theory to understand the actions of suicide bombers, for example, is to view their acts of terrorism as a choice for which the various competing options have differing costs and rewards – collectively known as *pay-offs*. If the perceived pay-off resulting from choice one is higher than for choice two, then the rational thing to do is to take that first course of action. The personal pay-off from killing oneself might arguably be characterised as zero – in the sense that the person dying by suicide will no longer be around to accrue costs or rewards. Suicidal individuals may characterise their day-to-day experiences as painful, saddling them with a cost instead of a reward – a negative pay-off. When viewed from this perspective, the zero personal pay-off provided by death can seem the only rational choice. Historically, many cultures have considered suicide to be a rational response to ill health, dishonour or other forms of suffering, although in modern Western society we are not typically of this view anymore.[80]

It is also important to be aware that pay-offs can be derived not just at an individual level, but can extend to family members or even friendship groups. For some species of animals (some biologists would argue all species), self-interest can extend even towards family members who do not yet exist. Animals that practise suicidal reproduction – a mating strategy in which one of the partners sacrifices their life – calculate carefully that the benefits they (or arguably more correctly their genes) gain through siring multiple offspring are worth dying for. Some male orb-weaver spiders, for example, sacrifice their lives upon inseminating a female. Their body remains attached to – and essentially blocks – the female's genital opening. This makes it harder for other males to subsequently mate with the female and thus increases the chances that the progeny she sires will share his genetic material.[81] Sacrificing his life to increase the expected number of off-spring he sires is a trade-off worth making for the male orb-weaver.[82]

For some groups of people, pay-offs can be derived from the impact of their choices on wider ideological or political organisations. For example, demonising the enemy offers a simple way to indoctrinate military personnel, creating and maintaining a shared sense of purpose and, consequently, encouraging actions which are for 'the greater good'. Soldiers can be further convinced to make 'the ultimate sacrifice' through promises of honour, fame and status, even if achieved posthumously. Political activists may deem that the societal reward from sacrificing their life for 'the cause' outweighs the personal benefit they continue to receive by staying alive. In each case, the expected pay-off that the person perceives to be gained from sacrificing their lives outweighs that which they associate with staying alive.

Either on its own or when mixed with a combination of political, ideological and military-type motivations, a belief in the infinite rewards promised in the afterlife of a 'martyr' might, for a prospective suicide bomber, outweigh any finite reward offered by living peacefully. Hamas recruiter Muhammad Abu Wardeh boasted of promising jewelled recliners, unlimited reserves of hangover-free wine, delicious food and, of course, the well-publicised seventy-two virgin wives, as incentives to would-be suicide bombers to join the infinite afterlife in paradise.

While from the outside it may appear that people making these extreme life-ending decisions might be acting irrationally, this is rarely the case. Very few, other than perhaps the very young and those suffering a small number of mental-health disorders, can genuinely be said to be irrational. Indeed, researchers investigating the motivations of suicide bombers who had attempted but failed in their mission found that the majority did not harbour psychological disorders and before their indoctrination by terrorist groups were relatively well-adjusted people.[83]

Unsurprisingly, the theory suggesting that we might take decisions which maximise our expected pay-off isn't a new one. In the seventeenth century, French mathematician Blaise Pascal even used the idea to argue for the rationality of believing in God. In a thought experiment now referred to as *Pascal's gambit*, he likened the decision of whether to believe in God or not to betting on the outcome of a coin flip. If God doesn't exist and you guess this correctly, then the pay-off might be that you enjoy your earthly pleasures slightly more than you would have done if you had lived a pious, God-fearing life, but nothing more. Conversely, if God really does exist and you put your money on this, then, Pascal argued, the sacrifices made in your mortal life to uphold this belief would be far outweighed by your infinite reward in heaven. So, he reasoned, no matter how small the probability that God exists, when multiplied by the infinite pay-off for this result coming true, the expected reward for believing in God will always outweigh the finite expected pay-off for not believing. The rational thing, Pascal concluded, was to believe in God.

Karl Marx argued that expected-pay-off arguments like this have been manipulated in many different societies. In suggesting religion was the opiate of the masses, he argued that the distraction of the promise of rewards in the afterlife supressed the masses' desire to fight against societal inequality to improve their lot in this life and instead to accept the status quo.

First-strike advantage

Now that we've couched the idea of rationality in terms of the expected pay-offs of cynical game players, we can return to analyse the example of the conflict in the Middle East. Imagine that we have just *two* opposing states (think Israel and Egypt in the scenario of the build-up to the Six-Day War). For simplicity, let's call them player A and player B in our simplified 'war game'. Imagine, instead of fighting over land, resources or strategic positions, the players in our simplified game have a total prize of £100 at stake, which will be divided up between the two of them. Let's say, for the sake of argument, that player A has a 60 per cent chance of winning if they end up fighting it out over the money, and player B has a 40 per cent chance.

The players then have a choice as to whether they fight it out (war) or negotiate a settlement. If both choose to negotiate, then they will try to come to an amicable agreement without fighting, but if either party chooses to fight, they both get dragged into the ensuing conflict. The problem with fighting it out is that it costs both sides.

Let's say, for the sake of argument, that they each incur a cost of £10 if they go to war, whereas negotiation, in contrast, is free. Given their chances of winning (60 per cent and 40 per cent, respectively), if the players decide to go to war, then A would expect to win £60 minus the £10 cost of the conflict, giving a total prize of £50. Player B would expect to gain £40 but expend £10 fighting and head home with £30.

What happens if, instead of fighting, both sides decide to negotiate? The lack of costs associated with negotiating (compared to going to war) means that there is a sweet spot that can be found at which both players head home with more money than if they'd ended up fighting each other. The £20 total cost of fighting can be distributed between the two parties instead of lost in the futility of war. In fact, this extra cash means that there is a whole range of possible outcomes for which an agreement is favourable for both sides. If A takes home anywhere between £50 and £70 of the total £100, they are better off than with the £50 they would have achieved through fighting it out, while B's corresponding settlement – between £50 and £30 – is also superior to the £30 they would otherwise be taking home. Perhaps the fairest settlement to both sides would be the 60–40 split which reflects their chances of winning the conflict they declined to enter into. The potential outcomes for the two players, given their individual choices of whether to negotiate or to fight, are shown in Figure 5-1.

For many potential disputes, a negotiated solution in which any spoils are divided pre-emptively, avoiding the costs of conflict, can be beneficial for both sides. Simple game theory says so. This is why mediation is almost always the go-to option for couples looking to divorce – rather than a messy, expensive, drawn-out court case, in which both sides stand to lose face (and money). The ability to actually achieve such peaceful resolution through negotiation when the players are stubborn human actors is a different matter.

PLAYER B PLAYER A	NEGOTIATE	FIGHT
NEGOTIATE	60 , 40	50 , 30
FIGHT	50 , 30	50 , 30

Figure 5-1: In game theory, we often describe the possible outcomes of games between two players through a pay-off matrix. The titles of the rows describe the actions ascribed to player A and those of the columns the actions of player B. In each of the four 'pay-off' boxes (corresponding to the four possible pairs of choices the two players can make), the first entry gives the expected outcome, or pay-off, for player A playing the game under those strategic choices. The second entry gives the outcome for player B. So, for example, when player A and player B both decide to negotiate, the top left box represents the settlement outcome of £60 for A and £40 for B. Since the expected pay-off for both players under a negotiation strategy is at least as good as under a fighting strategy (irrespective of what the other player chooses to do), this should be the strategy that rational actors choose to take.

Would Israel and Egypt both have been better off suing for peace, even after Nasser took the seemingly irreversible step of blockading the Straits of Tiran? Well, quite probably not. One important factor that the simple game we outlined above fails to take into account is that it can make a difference who plays their hand first.

In the overly simplified game above, deciding to negotiate when your opponent wants to fight doesn't put you at a disadvantage. If you choose to fight, irrespective of your opponent's plans, you never do any better than if you had first chosen to negotiate. In the game defined above, the negotiation strategy is no worse than the fighting strategy, even if your opponent opts for conflict. As long as at least one party chooses to fight, the other party gets dragged in and the outcome is the same whether they initially tried to negotiate or not.

In reality, though, if one state decides to initiate a war, while the other is unprepared or naïvely hoping to head to the negotiating table, this can lead to a significant *first-strike advantage*. A party that chooses to strike first would carry with them the element of surprise, as well as dictating when and how the early fighting takes place. If the potential first-strike advantage is big enough compared to the costs of fighting, this can change the game so that war becomes inevitable even before the first blows are struck – mutual negotiation is always off the table.

To mimic this scenario, imagine now that in the £100 prize game we considered above, catching your opponent unawares by striking first gains you an advantage of £30, while costing your opponent £5 in addition to the £10 cost of war that you both still incur. The four potential outcomes of this two-player game are illustrated in Figure 5-2.

PLAYER A / PLAYER B	NEGOTIATE	FIGHT
NEGOTIATE	? , ?	45 , 60
FIGHT	80 , 25	50 , 30

Figure 5-2: The pay-off matrix for the game in which one player chooses to strike while the other is still vacillating. There is no negotiated settlement which will see both players improve upon the outcomes in which they strike first.

If A chooses to strike first, while B offers to parley, then A's expected outcome is a 60 per cent share of the £100, plus £30 for going first, minus the £10 costs of war, giving an £80 net pay-off. B's expected pay-off, comprising the 40 per cent share of the £100, less the £10 cost of conflict and the £5 penalty for being late off the mark, leaves them with just £25. With the roles of negotiator and aggressor reversed, B takes home £60, while A receives only £45. If both come out all guns blazing, fully prepared for the fight, then they receive the same £50–£30 split as the conflict scenarios in the first game. There is no negotiated strategy which will see both A *and* B do better than they would do by striking first. It's impossible to divide even the full £100 prize to give A more than £80 and B more than £60 (their pay-offs for striking first) simultaneously. When the net advantage from striking first (£35 in this case, comprising the £30 bonus and the opponent's £5 penalty) exceeds the total cost of war (£20 in this case – £10 for each side), the possibility of negotiated settlement goes out of the window. If either side realises this, it makes strategic sense for them to act first and conflict becomes inevitable.

It is also possible that the advantage for striking first and the cost of being dragged into a conflict unprepared are not equal for both sides. At the eleventh hour, when preparing to strike first, Egypt calculated, based on the potential withdrawal of Soviet backing, that the benefit of doing so was no longer as advantageous as they had thought. Israel, on the other hand, receiving no strong external censure, and understanding how decisive an early attack might be, had no such qualms. They figured that the benefits of striking first were so significant that they couldn't afford not to. And so it proved when they executed their plan. The fact that they delivered a humiliating defeat in just six days bears testament to the huge advantage they received by catching their adversary off their guard.

Although it imparted a huge benefit to Israel in the Six-Day War, first-strike advantage is rarely so beneficial; the expected costs of long-lasting, large-scale conflicts will typically outweigh any first-strike advantage. In the Second World War, Nazi Germany's blitzkrieg strategy generated a significant first-strike advantage, allowing them to invade and conquer Poland at lightning speed and almost unopposed. But the eventual cost of the war that dragged on for almost six years was easily enough to outweigh even the seemingly significant advantage that Germany gained by striking first.

Amoral dilemma

That large-scale conflicts ever occur seems almost perverse, given that mutual negotiation holds the potential to bring about a better resolution for both parties than mutual fighting. Even when a first-strike advantage exists, the situation in which both sides sit down at the negotiating table will avoid the cost of war. This has the potential to provide both sides with a bigger benefit than if they both end up being dragged in. In fact, such perverse outcomes are well known in game theory. The first-strike-advantage scenario which describes the build-up to the Six-Day War is mathematically equivalent to perhaps the most famous game in all of game theory – the prisoner's dilemma.

The game is usually described in terms of the options facing two members of a renowned criminal gang – let's call them Amy and Ben – being held incommunicado at the police station. The police know they can pin a minor charge on both suspects, which would see them each do a year in jail. Naturally, however, they would prefer to nail the pair for the much bigger crime they are believed to have committed, but for which there is insufficient evidence to convict either. To elicit a confession, the police offer both prisoners a deal. If Amy confesses to the crime and implicates Ben, while Ben stays silent, then even the minor charge against Amy will be waived in return for her cooperation. This leaves Ben to take the full blame for the crime, resulting in a sentence of ten years. Ben is offered the same deal to rat on Amy. The kicker to the deals is that if both prisoners grass each other up, then they both go down for the bigger crime, serving five years each.

AMY \ BEN	COOPERATE	DEFECT
COOPERATE	1 , 1	10 , 0
DEFECT	0 , 10	5 , 5

Figure 5-3: A pay-off table for the prisoner's dilemma game. Although, in our scenario, both prisoners cooperating with each other by keeping quiet would lead to a better outcome for them than both defecting, the temptation to rat out their co-conspirator is too great to resist.

The pay-off table for the different combinations of strategies adopted by Amy and Ben for this prisoner's dilemma is shown in Figure 5-3. We refer to the strategy of keeping shtum as a prisoner attempting to *cooperate* with the other, while ratting the other prisoner out is termed *defecting* from the deal. While a strategy of mutual cooperation would serve both parties better than one of mutual defection, this strategy doesn't make sense for either party to stick to, based purely on the consideration of prison time alone.

Let's look at the situation from the point of view of Amy. If Ben chooses to cooperate, then, by choosing to defect, Amy can walk free instead of serving a year-long sentence. Defecting is the rational choice. Alternatively, if Ben has chosen to defect, then Amy reduces her time from ten years down to five by defecting. Whatever Ben's choice, Amy's best response is always to defect and turn informer. Acting rationally, Ben is thinking exactly the same thing and knows that his best choice is also to defect. Despite bilateral cooperation offering a mutually beneficial arrangement, the only rational solution to the game of prisoner's dilemma is mutual defection.

This was christened an *equilibrium* by legendary game theorist John Nash, whose life is depicted in the wonderful mathematical biopic *A Beautiful Mind* – namely, a set of strategies that each player is incentivised to take and, if acting rationally, will not deviate from. At the Nash equilibrium, no player has anything to gain from changing their strategy and, in the brutally cynical word of the game theorist, if the proposed action is not in a player's best interests, they don't do it. The only rational solution to the prisoner's dilemma is mutual backstabbing.

That's not to say there aren't games whose resolutions – *Nash equilibria* – are beneficial for all the players. If the pay-offs for the different strategies change, that may fundamentally alter the type of game being played. With the prisoners' scenario described above, there may be other benefits or costs which might stabilise a mutual-cooperation strategy. Known criminal informants tend to be received badly both inside and outside prison, making defection more costly. Cooperation might be incentivised by rewards for keeping shtum, which would change the pay-offs in favour of mutual cooperation. Indeed, the very evolution of complex life on earth depends on finding solutions to this problem – ways to change the pay-off matrix (or change the game more radically) – so that cooperation is more beneficial to individuals than defection.

The road not taken

To determine the best strategy to play in any given situation, we really need to know how the people we are up against will act. And it isn't enough to simply ask our opponents what they plan to do in a given situation because, in the real world, people lie. They say they will do one thing and then go off and do another. The job of the game theorist is to see through the subterfuge to try to understand what their opposition's true cost–benefit trade-offs look like. We can then use that understanding to predict what they will do in the future and, consequently, what our own best course of action should be.

This is where things get really difficult. Top poker players observe their opponents' actions to understand their strategy, so they might know better how to play against them. As an untrained observer, we might be impressed by the final flourish as one expert player throws down a royal flush and claims the pot. This is a piece of unavoidable truth the winner had to disclose in order to take the money. We learn something from that. But poker players very rarely tell the truth unless they have to. Most of the time they are lying – either by omission or by unabashed bluffing. The cards thrown face up on the table are useful, but we would learn far more from the hands we never get to see – the bluffs that force people with better hands out of the round and the folds that demonstrate that the expert knows when to cut their losses.

In trying to draw on information from the past to infer rules which govern the future, it is a common mistake to construct cause-and-effect narratives based solely on the events we observe. We are impressed by the poker player's exploitation of their good hand to win the big pot. We infer that to win at poker, you have to be dealt good cards. While this is true in part, we forget the fact that the poker expert was only able to remain in the game to play the good hand because, unspectacularly, she knew when to fold in earlier rounds and to not go broke chasing bad odds. Poker is at least as much about the hands you don't play as the ones you do. Often, the absence of evidence can tell us something useful.

It is the job of the historian to suggest why one event caused another to occur and to draw lessons from those incidents which appear in the historical record. Much of our perspective on past events then comes from this causal reasoning about the events that actually happened. In contrast, it is the job of the rational game theorist to fill in the gaps – to think about the roads not taken and the what-ifs, the counterfactuals and the events which did not occur and, importantly, the reasons they did not. Much of the retrospective narrative of warfare, for example, is subject to this told/untold bias. Few textbooks concentrate on the many crises that are averted, focusing, rather, on the handful that were not. Despite school history lessons seeming to jump from one war to the next, major conflicts are, in fact, unusual occurrences. As we saw earlier in the chapter, the costs associated with wars mean that the odds of the game are stacked against their outbreak. Sometimes, the stories that dramatically alter the course of history are the ones that never happened at all.

Stanislav Petrov. Does the name ring a bell? For most people it won't, but his actions, or rather his inactions, are so world-changingly consequential that he should really be a household name.

In the autumn of 1983, tensions between the US and the Soviet Union were running high. The late 1970s had seen a renewal of Cold-War hostilities between the two states after a period of relative calm. In 1980, the United States, under President Carter, had boycotted the Olympic Games in Moscow in protest at the Soviet invasion of Afghanistan in 1979. When Ronald Reagan came to power in late 1980, he decided not to engage in nuclear-disarmament talks with the Soviet Union, instead declaring that his strategy to end the Cold War was 'that we win and they lose'.

In 1981, the Bulletin of the Atomic Scientists' 'Doomsday clock' – its metaphorical hands placed symbolically close to midnight to represent the human race's proximity to nuclear catastrophe – had been set to 23:56. This was the closest it had been to midnight since 1960, closer even than during the Cuban missile crisis. The clock would be put forwards again, a further minute closer to midnight, in 1984. Throughout 1982 and 1983 the US conducted *psychological operations* – effectively high-stakes bluffing, on an international scale – in which US bombers would be flown directly towards Soviet airspace only to deviate from their paths at the last minute.

On 1 September 1983, a civilian plane, Korean Airlines flight 007 from New York City to Seoul, accidentally strayed into Soviet airspace. It was rapidly brought down by a Soviet interceptor aircraft. Larry McDonald, a member of the House of Representatives for the state of Georgia, lost his life in the incident, along with 268 other passengers and crew. Although the Soviet Union initially denied all knowledge, the United States used the incident to galvanise support among NATO allies for the installation of Pershing II and Gryphon cruise-missile systems in West Germany. In response, the Soviet Union upped their missile-surveillance capabilities. They were

primed to expect a nuclear attack from the United States and geared
to retaliate before the missiles made landfall.

It was against this background that Lieutenant Colonel Stanislav
Petrov of the Soviet Air Defence Forces sat in the secret Serpukhov-15
bunker, home to the command centre of the Soviets' Oko nuclear
early-warning system. It was explicit Soviet policy that, if the early-
warning system detected inbound missiles, a retaliatory nuclear strike
against the United States should be launched immediately to bring
about the mutually assured destruction of both superpowers. In the
event of such an attack, it was Petrov's job to pass the message up the
chain of command in order to trigger the retaliatory strike.

Just after midnight on 26 September 1983, Petrov sat at his desk,
languidly monitoring readouts from the various computer screens
in front of him. Suddenly, a siren sounded, abruptly interrupting his
inertia and rousing him to attention. As adrenaline coursed through
his veins, one of the backlit screens in front of him flashed up the
word 'LAUNCH' in huge bold red letters. As abruptly as it had
started, the siren stopped, but the screen continued to flash up its
warning, indicating that missiles had been launched by the United
States. Without any hyperbole, Petrov was faced with one of the
most important single decisions in the history of the world. Should
he escalate the warning up the chain of command to his superiors or
sit on it, hoping it was a false alarm? If he passed the news on up the
line to the top, he was certain that his commanders would accept his
judgment and would act to launch a retaliatory strike.

As he continued to contemplate the magnitude of the responsibility on his shoulders, the siren went off again. A second missile
launch had been detected, quickly followed by warnings of a third,
a fourth and a fifth. The pressure was mounting on Petrov. He had

to act soon. Every second he wasted would reduce the time afforded to his superiors to launch their strike. If he waited too long, the matter would be out of their hands. In front of his eyes, the computer screens suddenly switched their message from 'LAUNCH' to 'MISSILE STRIKE'.

Still, Petrov continued to sit on his hands. Something didn't quite add up. He reasoned that a genuine nuclear first strike from the US would not comprise one or two or even a handful of sequentially fired missiles, but hundreds of launches and all at the same time. A few rockets appearing one at a time didn't make sense. A couple of minutes after the first warning, ground radar had still failed to pick up a corroborative signal. This only served to increase Petrov's distrust in the reliability of the novel launch-detection system, only recently deployed by the Soviets. He was still in two minds as to whether to trust the message flashing on the computer screens in front of him, impelling him to act. After a few more moments' hesitation, he decided he could wait no longer. He picked up the phone line and rang the number of the officer on duty at army headquarters. 'Colonel,' he said to his superior, 'I have to report . . . a system malfunction.'

Even after he hung up the phone, he wasn't sure he had made the right call. He knew that it would, by then, be too late for anyone to correct his error, so he waited. When, after twenty-five minutes, he was still sitting unmolested at his desk, he became confident that he had been right. The warnings had been false alarms.

In actuality, the Soviet satellite, primed to detect the flare of an intercontinental ballistic-missile tail, had picked up sunlight reflecting off the high-altitude clouds above North Dakota. Petrov's cool-headed actions that night almost certainly helped to ward off an unprecedented nuclear war.

MAD world

Interestingly, knowing about Petrov's inaction tells us something important about the game that was playing out between the USA and the USSR. A favourite trope of historians is that the mutual stock-piling of weapons increases the chances of a war breaking out. This argument is based on the fact that some high-profile wars, notably the First World War, have been directly preceded by such arms races. However, making this argument is to ignore all the situations in which the build-up of arms was not followed by the outbreak of hostilities. The most famous example is, of course, that the Cold War never became hot, despite numerous opportunities. And the reason it stayed cold was precisely because of the nuclear stockpiles harboured by both sides. The cost of tensions bubbling over would have been far too great.

The acronym MAD, standing for Mutually Assured Destruction, describes the situation in which the build-up of arms on both sides acts as a deterrent to either one ever using those arms. Soon after the end of the Second World War, both superpowers had stockpiled sufficiently destructive nuclear arsenals that the threat of total annihi-lation if either side were to launch an attack became a near certainty. Ironically, this threat is enough to ensure that neither side has ever used their weapons.

We can think of the two superpowers' choices in terms of a two-player game. At any moment in time, the USSR and the USA are both deciding whether to launch a nuclear strike or whether to hold fast. The pay-off matrix for the game looks something like Figure 5-4. If both parties continue to hold, then each receives a negative pay-off of -100, reflecting the expense of building and maintaining a nuclear arsenal, as well as the uncomfortable tension surrounding the MAD-mediated peace. This cost pales into insignificance, however, compared to the pay-off for launching an attack. Choosing to launch effectively ensures that both parties will be wiped off the face of the earth, which is represented by a pay-off of $-\infty$ – and nothing could be worse for either party than that.

USSR \ USA	HOLD	LAUNCH
HOLD	-100 , -100	$-\infty$, $-\infty$
LAUNCH	$-\infty$, $-\infty$	$-\infty$, $-\infty$

Figure 5-4: The pay-off table for mutually assured destruction (MAD). If both sides choose simultaneously to hold off from attacking, they receive a small negative pay-off. The consequences of choosing to launch, however, are always infinitely bad.

To understand what will happen in the game we need to think about how each superpower best responds to the other's actions. If, at some moment, the USSR is choosing to hold, then the US' best strategy would also be to hold. If, however, the USSR decides to launch a nuclear missile, it doesn't matter whether the US was deciding to hold or to launch at the same moment. If they were choosing to hold, then as soon as their early-warning system notified them of the launch, they would fire their own weapons and the result will be mutually assured destruction (MAD).

In the prisoner's dilemma game, the best possible action for an individual is always to defect, independent of what the other party does. Here, in contrast, the best choice of action varies depending on what your opponent chooses to do. If your opponent chooses to hold, then you should also hold, but if your opponent is going to launch, you should launch, too. One side launching while the other side is holding isn't going to be a valid solution – or Nash equilibrium – because the best possible outcome if one side is holding is for the other side also to hold.

Given that hold–hold and launch–launch are both possible viable Nash equilibria for the game, it seems reasonable to assume that the strategy that will be arrived at will be hold–hold, which corresponds to world peace, rather than launch–launch, which brings about the end of days. But it's important that launch–launch is a credible solution. If either side believed that the other one would not retaliate when fired upon, then their incentive to stay their own hand would suddenly disappear. Ironically, the threat of complete destruction is essential to maintaining peace.

During the Cold War, both parties understood the threat of mutually assured destruction and played the game accordingly. The deterrent strategy is so effective that there hasn't been a war between two major world powers since the Second World War ended in 1945. Peace between the two superpowers was the sensible solution for two rational actors. But what would happen if one side were able to convince the other that they were not acting rationally?

MADman

In Chapter 3, we learned about the Naskapi people and their use of randomness to help avoid falling into hunting routines. It would have been tempting for the hunters to reason that previous success in one region of their territory would lead to successes in the future in the same area. But always hunting in the same place leads to predictability, which the other players in the game (in this case, the animals they were hunting) could exploit to avoid being caught. Game theoreticians might refer to the practice of choosing probabilistically from among a range of strategies to avoid becoming predictable as a *mixed strategy*.

As we saw in Chapter 3, players of rock paper scissors know that if you throw the same shape every time, you get picked off easily by your opponent. Regularly alternating patterns also get spotted quickly and are easy to beat. The best strategy, if you can't guess your opponent's move, is to choose completely at random between the three different plays. However, as we have learned already, for our pattern-trained brains, being truly random is easier said than done.

In football, penalty takers might use the mixed strategy of aiming at different parts of the goal to stop goalkeepers from figuring out where they will shoot. Indeed, a 2002 study found that, rather than consistently favouring one side of the goal, penalty takers in two of Europe's top leagues chose randomly between kicking to the left, the right or down the middle.[84] Remember, though, that this is not the same as simply alternating sides, which is an entirely non-random and easily predictable strategy.

In recent experiments into the impacts of emotional unpredictability,[85] management students were asked to negotiate a hypothetical venture with each other according to some pre-specified rules. In one scenario, negotiators were asked to be relentlessly negative and angry, while in another, they were asked to frequently change their emotional tone between positive and negative. The students whose counterparts displayed emotional unpredictability were made to feel as though they lacked control over the negotiations, leading them to make larger concessions and irresolute demands.[86]

In the context of international diplomacy, sticking to a pure strategy – having a preordained response for any given situation – might reduce the ability of a negotiator to bluff, bluster or manipulate an opponent. Conversely, when negotiating with a despot who is employing a mixed strategy – someone who might, for example, have their finger on the nuclear button one minute, while advocating for total disarmament the next – an opponent might find themselves making more concessions than they would to an actor whose rational actions they find easy to predict. One particular mixed strategy, a form of brinkmanship known in political science as the *Madman Theory*, was the basis of much of Richard Nixon's foreign policy in the late 1960s and early 1970s. The aim, as the name would suggest, was to convince Nixon's communist opponents that he was more than a little unhinged. He reasoned that if his opponents judged him to be an irrational actor, they would not be able to predict his plays and would thus have to make more concessions to avoid the risk of accidentally triggering him into retaliation.

In October 1969, with negotiations to end the war in Vietnam at a stalemate, Nixon put his Madman policy into action. In a calculated act designed to convince the Soviets of his recklessness and coerce them into exerting their influence over Hanoi, the president placed the US military on full global war readiness alert. On 27 October, unbeknown to the American people, Nixon launched operation Giant Lance. Eighteen B-52 bombers loaded with some of the world's most powerful thermonuclear weapons were dispatched at 500 miles per hour across Alaska towards the Bering Strait. Nixon knew that Soviet radars would pick the bombers up early on in their trajectory and see the threat coming. He hoped that this would scare the USSR into acquiescing to his Vietnam demands.

At the same time, although Nixon was unaware of the situation, the Soviet Union was engaged in a secret border dispute with China that had been rumbling for six months. The Soviets were wary of the Americans' newly found sympathies towards Beijing, the US having thawed previously frozen trade relations between the two countries earlier in the year. There was a very real danger that the Soviets might have interpreted Nixon's behaviour not as the irrational actions of an unbalanced president, but as the rational and strategic pre-emptive actions of a Chinese ally.

As the bombers approached the eastern border of the Soviet Union, they slowed and diverted their course, so as not to stray into Soviet airspace. Soviet General Secretary, Leonid Brezhnev, was concerned enough by Nixon's Madman antics to demand that his ambassador set up an urgent meeting. The B-52s spent three more days harrying the border before being ordered home in what Nixon hoped was another unpredictable move, demonstrating that he could de-escalate tensions as quickly as he had heightened them. Although the operation may have initially spooked the Soviets, Nixon's game-theoretic gamble was ultimately a miscalculation, needlessly escalating the risk of nuclear war, failing to bring Hanoi back to the negotiating table and leaving him in a weakened position after capitulating in his game of nuclear chicken.

It's a truel world

In exemplifying the power of game theory to predict and even control the future we have, for simplicity, boiled all the situations we've considered down to only two-player games. In reality, however, games might be played between more than just two parties. Indeed, one of the most studied scenarios in game theory involves a confrontation known as the truel – the three-person equivalent of a duel. Unlike the duel – in which we would expect the person with the best accuracy and fastest draw to have the greatest probability of winning – the addition of a third shooter can sometimes lead to surprising results. But the outcome depends heavily on the rules of the game.

Truels are a popular trope in the cinema, having been used to resolve plot issues in at least three Quentin Tarantino movies alone. Probably the best-known example, though, features in one of the most famous movie scenes of all time. Towards the climax of *The Good, the Bad and the Ugly,* the three eponymous characters are standing in a triangle on the perimeter of a circular plaza each with hands hovering around their waists ready to draw. We'll use this set-up to describe the situation in our own truels.

Imagine that our truellers – Good, Bad and Ugly – are standing roughly equidistant from each other. We'll make the very reasonable assumption that each of them wants to maximise their chances of survival. In the first scenario, let's consider that the players take it in turns to fire in the order, Ugly, Bad, Good, that they are perfect shots, but that each only has one bullet. What should Ugly do to maximise his chances of survival?

If Ugly shoots and kills Bad, then it will be Good's go next and

Good will kill Ugly. Ugly reaps the same reward if he aims at and kills Good, rather than Bad. Going first is not to Ugly's advantage and seemingly always ends up in his death, unless he can think his way out of the situation laterally. The counterintuitive solution to Ugly's problem of having to lead out is for him not to aim at either Bad or Good, but to fire into the air. That way he eliminates himself as a possible threat. Bad will then go on to shoot Good and both Bad and Ugly will survive.

In a slightly more realistic scenario, let's imagine now that the gunslingers are all of differing abilities and each has unlimited ammunition, so that the truel goes on until there is only one survivor. Let's assume Good is the best shot, hitting his target nine times out of ten. Bad is next best, hitting his target seven times out of ten, and Ugly is the worst, hitting his target only half the time. To even the odds, the gunslingers have agreed to take shots in turns in order of ability from worst to best, until only one person remains standing. How does the gunfight play out in this revised scenario?

As is so often demanded by game theory, we must to put ourselves in the gunslingers' shoes. It makes sense for each shooter to be most concerned about the competitor with the best aim. Good should target Bad and Bad should target Good. Ugly could also target Good to try to take the best shooter out of the game. If they proceed with those strategies, then, ironically, the best shooter in the pack, Good, with two people shooting at him from the off, wins with only a 6.5 per cent probability. The second-best shooter, Bad, wins 56.4 per cent of the time and the worst shot, Ugly, wins with a surprisingly high 37.1 per cent probability. Being known to be the best shot can be a significant disadvantage in the truel, as it can cause the others to gang up on you.

In fact, though, learning from our previous out-of-the-box solution, we find that Ugly can do even better by consistently firing into the air in the early stages. Since Good and Bad will be more concerned about taking each other out, Ugly 'wasting' his shot effectively reduces the truel to a duel between the two better shots. As soon as one of them takes the other out, Ugly will get the first shot in the duel between himself and the survivor. This first-shot advantage in the resultant duel swings the balance in Ugly's favour, outlasting the two better shots 57.1 per cent of the time (compared to 37.1 per cent when actively firing from the start). Now Good wins slightly more frequently at 13.1 per cent of the time (up from 6.5 per cent previously) – still disadvantaged by going last, but at least no longer targeted by Ugly – and Bad's probability of winning is reduced to 29.7 per cent (from 56.4 per cent), as he no longer has the support of Ugly to take out Good. In fact, even if they draw lots to see who goes first rather than letting the worst shot go first, Ugly is still the most likely winner if he shoots into the air until one of Good or Bad takes each other out. It seems crazy that the best shot can be most likely to lose, and the worst most likely to win, but it's not just an artefact of this artificial gunfight.

In the run up to the June 2009 Virginia Democratic gubernatorial primaries, state senator Creigh Deeds was floundering. In one January poll he registered just 11 per cent support. Over the next four months, he only polled higher than 22 per cent once, as the other two candidates, Terry McAuliffe and Brian Moran, swapped the polling lead between themselves. Deeds' fundraising campaign was also stuttering. In the first quarter of 2009 – a crucial period ahead of the election – he had raised just $600,000 compared to Moran's $800,000 and McAuliffe's $4.2 million. But in mid-May, the game suddenly changed.

The candidates began to plough much of their remaining resources into advertising. Moran went hard at his main rival, McAuliffe, criticising his record as a businessman. McAuliffe responded to his biggest threat, Moran, with his own ad, defending his record and accusing Moran of 'trying to divide Democrats'. Moran hit out again, criticising McAuliffe's campaign against incumbent president, Barack Obama, in the Democratic primaries preceding the 2008 election. Moran hoped that this would diminish McAuliffe's standing in the eyes of the state's crucial African–American voters. All the while, as the top two candidates chipped away at each other's reputations, unassuming underdog, Creigh Deeds, was planting seeds of positivity with his self-promoting advertising campaign. When the *Washington Post* came out and endorsed Deeds in late May, many undecided voters recognised him as a reasonable alternative to the two former front-runners. Deeds' popularity in the polls shot up and by early June he was polling at over 40 per cent. Each of the formerly stronger rivals seemed to have managed to convince Virginian voters that the other was not electable. In elections on 8 June, Deeds won just under 50 per cent of the vote to McAuliffe's 26 per cent and Moran's 24 per cent – a landslide for the weakest candidate.

Allowing a truel to develop may have done the Democrats a disservice. In a two-horse race, declines in popularity for one candidate are equivalent to gains in popularity for the other. If it is harder to boost one's own image than it is to denigrate the other candidate, then the incentive is for the candidates to batter each other with negative advertising, leaving the electorate to choose between a rock and a hard place. In the truel, however, when negative-advertising campaigns diminish the reputations of the stronger two candidates, this can allow the weakest one to prevail. Deeds would go on to lose the resulting gubernatorial election to Republican candidate Bob McDonnell, while McAuliffe, the strongest of the three Democrat candidates on paper, would eventually go on to become Governor of Virginia in the election four years later. The Democrats may have served their purpose better if they had allowed just two candidates to do battle with each other in the primary.

Away from politics, a favourable strategy for weaker participants in a multiplayer competitive game – namely, staying in the background while the best fighters duel it out – has been arrived at naturally, over and over again, in the animal kingdom. For many animals, there are just two simple aims in life: to survive as long as possible and to reproduce as much as possible. In fact, through the lens of evolutionary success, the first goal just serves effectively to increase the possibility of the second – longer-lived animals having more opportunities to reproduce. The really important thing for most animals is to ensure their genetic material is passed on to the next generation. In many species, female parents invest more in each of their offspring than males, which means that females benefit from choosing a mate carefully. Conversely males, who invest relatively little in each offspring, benefit from reproducing with as many females as possible. Consequently, males often compete to mate with a female, to protect or win a harem, or for the best chances of fertilising a female's eggs, depending on the species. In many species this means males fighting with rivals, sometimes to the death. It's estimated that up to 6 per cent of rutting red deer stags are permanently injured each year, and that many of these will even die of their injuries.[87]

While two of the most impressive specimens fight it out, killing or injuring each other, subordinate males can nip in and mate with the female. So well established is this practice across the animal kingdom that it has its own name. *Kleptogamy* is derived from the Greek words *klepto*, meaning 'to steal' and *gamos*, meaning 'marriage' or, more literally, 'fertilisation'. Natural selection suggests that if only the alpha males were reproducing, then the variation in male fitness in future generations would become limited. The evolutionary game theorist John Maynard Smith came up with the theoretical idea of kleptogamy to explain how a wide range of male fitnesses could be sustained over time, although he and his colleagues preferred to call it the 'Sneaky Fucker' strategy.[88] And in some species, the evidence is there to support his hypothesis. A study of the mating habits of grey seals on Sable Island, off the coast of Canada, found that 36 per cent of females guarded by an alpha male were, in fact, fertilised by non-alpha males.[89]

In real gunfights, gunslingers don't stand around waiting to see the outcome of their opponents' efforts at the risk of being shot themselves. Instead, all three truellers must make up their minds about who to shoot beforehand and, in effect, play their hands simultaneously. This changes the dynamics of the game significantly, but can also lead to unexpected consequences, depending on the specific rules of the game.

In the game show *The Weakest Link*, contestants individually attempt to answer general-knowledge questions. Through multiple successive rounds, correct answers are rewarded with money, which can be banked into a collective pot for the team. At the end of each round, the contestants all write down who is the 'weakest link' – the person they would like to vote off the show before the next round begins. The contestants then reveal their choices simultaneously. The person who receives the most votes each round is eliminated.

At the end of the penultimate round, which features three contestants, we find ourselves watching a simultaneous truel playing out at the voting stage. To understand the dynamics of the voting in the penultimate round, like good game theorists, we must consider what comes next. Once there are just two players left, they first collaboratively add to the jackpot before going head to head. In the head-to-head each contestant attempts to answer more questions than the other and thereby win the total jackpot. Under these conditions, what is the optimal strategy for each of the three players to employ when voting in the penultimate round?

If the players are only interested in taking home the prize, irrespective of its final size, then each should try to play the strategy which sees them fighting it out with the weakest possible opponent in the final round. Contrary to the host's instructions to vote for 'who you think is the weakest link', each player should try to vote off the strongest competitor. As we have seen before, this can mean that the contestant with the reputation for being the strongest can be at a disadvantage. Both of the two weaker players should vote her off in order to end up duelling against the weaker alternative.

As a case in point, in a 2001 episode of *The Weakest Link*, contestant Chris Hughes (who had previously won both the TV quiz shows *Mastermind* and *Brain of Britain*) made it all the way to the end of the penultimate round without getting a single question wrong. He was by far the strongest competitor in the game, having been the strongest link – answering the most questions correctly – in each of the six previous rounds, and his opponents knew it.

As it transpired, being known to be the strongest player was, indeed, his downfall. The two weaker opponents voted tactically to eliminate him. In bidding him farewell from the show, host Anne

Robinson at least spared him the indignity of suffering her customary dismissal, 'You are the weakest link, goodbye'. Instead, she summed up Chris' predicament as follows: 'Chris, you have failed to answer any question incorrectly. You are the best contestant we have had on *The Weakest Link*, but you are too good for Seamus and Mari. With two votes they have voted you off – goodbye, Chris.' In the final round, Seamus, who had answered the fewest correct questions of the three in the preceding rounds, walked away with all the money. He may not have had the best general knowledge, but he did apparently have a decent grip on game theory.

In most editions of the game, however, it is either not sufficiently obvious who is the strongest remaining contestant, or the players are not savvy enough to vote tactically. One study which analysed almost 400 episodes of *The Weakest Link*, from the UK, the US and France, found that the strongest predictor of the contestants' voting patterns was not motivated by tactical considerations (the vote which optimised their chances of winning) but rather by retribution – the vote which allowed them to exact revenge on players who had voted against them in previous rounds.[90]

Common tragedies

If the simple competitive truel, with just three players, gave some counterintuitive results, you can imagine how much more complicated the situation can become when there are even more players in the game. In global, national and local contexts, it is not hard to think of examples of 'games' in which multiple participants compete with each other over access to some shared resource.

Fishing is a classic example of such a multiplayer competitive game. Italian explorer Giovanni Caboto reached the area of ocean now known as the Grand Banks of Newfoundland in 1497. He was amazed that the seas were so dense with life that fish could be caught by simply dropping a basket into the water and raising it rapidly. Sixteenth-century Basque fishermen enthused that man could meta-phorically 'walk across the sea on the backs of the cod', so dense were the oceans with life. When the English arrived on the scene in the early 1600s, they saw cod so abundant near the shore that they reported hyperbolically 'we have hardly been able to row a boat through them'.

For centuries after these reports, the supply of cod seemed, to all intents and purposes, unlimited. A sure path to fortune in those heady days off the east coast of Northern America was to become a fisherman, or better still, a fishing magnate. For over 400 years, the seas just kept giving. In the late 1960s, sonar shoal detection and the advent of huge freezer factory trawlers coming from all over the world meant that the removal of fish from that area of the ocean became more efficient than ever. Hauls continued to rise year on year. In the fifteen years between 1960 and 1975, around 8 million tonnes of cod were harvested from the Grand Banks – the same as the total catch from the 200-year period between 1600 and 1800. In 1968, a record 800,000 tonnes of fish were taken out of these fertile waters,[91] but could these yields be maintained indefinitely?

The answer, as the fishing industry found out to its cost, was a resounding no. Fewer than five years later, the same fishing inten-sity was bringing in less than half the yields. Feeling that its natural resources were being unfairly exploited, the Canadian government acted to limit who could fish in the waters up to 200 miles off the country's coast. The international industrial-scale fishing boats were sent packing, but now the local Canadian fisherman wanted their slice of the pie. Scientists urged the government to act cautiously by setting low quotas that would allow fish stocks to recover. Their pleas were ignored over fears that quotas would destroy jobs in the fishing industry in the short term. Instead, Canadian fishermen were allowed to build their own factory trawlers and to fish the depleted seas with renewed intensity. Yields remained roughly stable throughout the 1980s, despite the increased harvesting effort – a sure sign of a fish population in decline.

Then, sure enough, yields almost completely collapsed in 1994. The year before the collapse, when scientists suggested extremely stringent quotas to give the populations a chance to recover, their advice was labelled as 'demented' by the Canadian Minister of Fisheries and Oceans. After the crash, the breeding population was estimated to be just 1 per cent of its level thirty years earlier. Despite previous reluctance, the Canadian government was now forced to implement a near-total ban on commercial fishing along the country's eastern coast; 45,000 people lost jobs either directly or indirectly related to the fishing industry. Decades on, cod stocks in the area have still not recovered.

*

The outcome of a multiplayer game involving a common finite resource – in this case, the fish in the ocean – which each party would like to exploit for their own benefit, is often referred to as a *tragedy of the commons*. It seems in the best short-term interest of individual players to exploit as much of the resource as possible. But if everyone acts in this way, it leads to the resource's depletion, so that eventually, everyone loses out. This is the prisoner's dilemma played out on a grand scale. Each party acting in what seems their own best interests in the short term leads to an outcome which is to everyone's detriment.

We are all guilty of it. Who has not spent time reading product reviews online before making a purchase, but then failed to leave a review of their own to help others navigate their way through the same choices? How many parents can say they never sent their children into school with a cold, rather than take the day off work to look after them, despite the fact it might inflict the same infection on other children and their families?

The tragedy of the commons causes us to question again what exactly we mean by rationality: acting rationally in the short term can be irrational in the long term, and rational actions at the individual level can lead to suboptimal outcomes at the level of the collective. This seeming dichotomy is neatly captured by the concept of *bounded rationality* – the idea that our choices are determined with reference to a certain information horizon, beyond which we cannot reason. The time allowed to make decisions, the inherent complexity of the problem we are faced with and the limitations of the human brain all contribute to the inability of seemingly rational people to always make the optimal choice.

We see the tragedy of the commons anywhere there is a shared resource which is open to exploitation by multiple parties – the crockery in a shared kitchen in a student house, for example. It's all too easy to leave your dishes in the sink unwashed. You gain a time advantage by doing this and avoid the unpleasant job of washing up. But if everyone acts the same way, then the clean crockery is soon used up and everyone loses out.

In healthcare, the tragedy of the commons has life-and-death consequences. The widespread prophylactic use of antibiotics to avoid infection in agricultural settings provides the short-term, individual benefit of protecting the farmer's livestock from disease and promoting the animals' growth. But the ever-present antibiotics place selective pressure on the disease-causing bacterial pathogens. Bacteria that, through random mutation, are better adapted to avoid antibiotics can grow and reproduce more quickly in this environment, quickly coming to dominate the population. It's then possible that they can disperse outside the setting in which they originally emerged, causing widespread antibiotic-resistant disease in both humans and animals.

Vaccination is another health intervention for which individuals, acting in what they believe is their own self-interest, can significantly undermine a shared goal. Vaccinating a sufficiently high proportion of the population can provide enough people with immunity that a disease can no longer gain a foothold in the group. Achieving this so-called *herd immunity* does not require everyone to have had the jab. A disease can be eliminated and people who can't have the vaccine – perhaps because they would have an adverse reaction – can be kept safe through herd immunity. Herd immunity here is the common resource that people exploit. Some people, perhaps concerned about the potential side effects or a perceived loss of bodily autonomy, will choose not to be vaccinated and rely on a sufficient number of others doing so instead. They can thus benefit from the population-level herd immunity without paying the perceived cost. Of course, if everyone thinks this way, we can never hope to achieve herd immunity, meaning that the disease can spread freely in the population. Everyone loses out. In the UK for example, at just 85 per cent, in 2022 measles vaccination is at its lowest rate for ten years – 10 percentage points lower than the 95-per-cent target required to harness the full benefits of herd immunity. This puts members of the UK's population at greater risk of contracting this unpleasant and deadly disease.

And there are greater existential threats even than unmitigated disease transmission. The pollution of water courses, the exploitation of fossil fuels, unregulated logging and the destruction of natural habitats through deforestation all present textbook examples of common tragedies. But arguably, the most important environmental common is the global atmosphere. By acting in our own short-term self-interest and failing to limit emissions of greenhouse gases, each nation in the world contributes to global warming. Sadly, if left unchecked, the resultant change in our climate will eventually lead to death and suffering on previously unimagined scales. Despite the global scientific consensus that we are on course for environmental catastrophe, the world's leaders are, so far, failing to see past the individual benefits of short-termism to strive for the long-term global solutions that we need to combat anthropogenic climate change.

Changing the game

Real-life situations that appear to encourage short-term individual gain to the long-term detriment of the collective don't always have to remain that way indefinitely. By thinking carefully about how we can restructure the game and its pay-offs, we can avoid these common tragedies. The key is to change the rules, so that cooperation becomes mutually beneficial – so that what is in the best long-term interest of the collective is also in the best short-term interest of the individual.

Perhaps counterintuitively, one possible way to achieve this is to privatise the commons. In the context of a literal area of common land (from which the phrase tragedy of the commons derives), this might mean giving over parcels of land for allotments or gardens. People typically spend significantly more time maintaining their own gardens, from which they derive some of the benefits of exclusive use, than they do looking after the local shared common. While its loss to privatisation might seem to defeat the point of a common resource, some of the global benefits might still be maintained. While the general public lose access to privatised gardens, for example, the maintenance and management of these outdoor spaces at the individual level might retain the pay-offs of improving habitat and species diversity, carbon capture and other environmental benefits, to the advantage of the collective.

This might work well for resources with definable boundaries like land, but not all common resources can be parcelled up and privatised in this way. In any case, such privatisation isn't always desirable. Regulation which limits the amount of a common resource available for each user can also solve the problem. The punishments for exceeding mining extraction limits, logging allowances or hunting quotas, for example, need to be sufficiently large and rigorously enforced that by overexploiting the common resource, parties stand to lose more than they gain. If the costs are high enough, then the exploitation of the common resource is no longer in the individual's best interest.

Alternatively, rewards for good behaviour can incentivise globally beneficial solutions. Cycling is a classic example. By providing improved infrastructure for cyclists, including dedicated bike lanes, secure parking facilities and priority traffic signals, Copenhagen has incentivised cycling over car driving. By making cycling the quickest, cheapest and safest way to get around the city, the mutual benefits of public health, reducing pollution and improved productivity can all be simultaneously protected. In 2009, 55 per cent of all vehicular journeys in Copenhagen were taken by bike. Part of Copenhagen's success in re-introducing bikes, in a city which was threatened with being overwhelmed by cars in the 1960s, is the social reinforcement of good practices. Brightly lit signs posted around the city flash up the number of fellow cyclists out on the roads that day, bolstering a sense of collective purpose in Copenhagen's citizens. The use of cars is not prohibited; it has just been turned into a more costly individual choice.

Such centrally regulated solutions are not always possible, how-ever, as they require global oversight. Better still are self-policing, locally enforced solutions. These are particularly apt in small commu-nities with strong social norms. If you live in a village where everyone knows everyone else, you may be less likely to litter, for example, for fear of the reprisals that accompany being caught breaking this social taboo.

There is a multitude of inventive ways that a game-theoretic per-spective can help to change our collective viewpoint in order that we may resolve population-level tragedy-of-the-commons problems. Perhaps unsurprisingly, there are also plenty of ways in which game theory can help us to reframe some everyday, individual-level prob-lems to our own advantage.

The car's the star

In his excellent book on game theory, *Predictioneer*, Bruce Bueno de Mesquita describes his strategy for buying a car. Instead of heading into a dealership, which he describes as a *costly signal* – letting the dealer know you are keen to buy the car from them and giving them the upper hand in negotiations – De Mesquita does all his bargaining over the phone.

He first decides on the exact spec of the machine he would like to buy: the make, the model, the colour the trim, etc., and then he identifies any dealers in the local area who are selling that specific vehicle. He checks out the guide price each dealership is offering and then he picks up the phone.

He rings the dealerships one by one and is completely upfront with them about what he is doing. He tells them of the car he would like to purchase and that he has identified the same vehicle at multiple dealerships in the area. He informs the sales agent that he will ring around each of the other dealerships to ask for their lowest price. Whoever gives him the cheapest quote will win the sale that afternoon, he tells them – 'So what is your best price?'

It's a brilliant strategy, which lets the dealer know simultaneously that you are serious about buying the car because you have done your research, but also puts the onus on them to treat you with honesty, to make their genuine best offer or risk losing out on the sale.

De Mesquita reports that sometimes the agents will complain that when he quotes their 'best price' to the next dealer, they will just undercut it and De Mesquita will then accept that lower price. 'That's right,' he tells them. 'So if you can go $50 lower, this is your opportunity.' And sometimes the dealers will refuse to quote a price on the phone, insisting that he makes the costly gesture of coming into the showroom to get the quote, limiting his buying options and putting him at the disadvantage in negotiations. De Mesquita sees this failure to quote a price as an admission on the part of the dealership that they can't compete with what others are offering and will simply walk away from that particular agent in the knowledge that he has several more on his list.

De Mesquita has bought at least ten cars this way, shaving thousands of dollars off the prices quoted on the internet. His formula has been replicated by his students and even journalists trying out his technique as non-experts in the real world.

Changing the typical rules of the negotiation is a win–win. The dealer benefits by saving time and energy, and not selling for less than they can afford, while the buyer benefits from the best possible price. Everyone is better off. Throughout the whole process, each dealer has the same chance to sell you their car, which they likely wouldn't have had if you'd just gone to one of them and tried to negotiate a price. Part of the key to changing the game is to recognise the possibility of mutually beneficial solutions that improve the outcome for all concerned.

Zero to hero

A zero-sum game is one in which one player's loss is another player's gain. Many sports and games, including chess, boxing, tennis and poker are zero-sum. The pot taken home by the winner in a private game of poker (as opposed to a game in a casino where the house may take a cut) equals the losses of those unfortunate or unskilful enough to go all in and lose.

The thrill of winning a zero-sum game is undeniable, but sometimes losing hurts, and in multiplayer games like poker, with only one winner, most people will walk away on the losing end. The zero-sum games I play with my kids almost invariably end in an argument or someone getting upset (often me!). The most enjoyable games we play as a family are collaborative ones, in which everyone can win by working together. All too often in real life, we naïvely assume that the 'games' we are playing are zero-sum, when, in fact, they aren't. There are many games in which one person's gains don't have to be balanced by another's losses – everyone can walk away happy.

Frisbees, the flying-disc toys, were first sold to the United States' general public in 1957. By the early 1960s, the craze had still not caught on in the UK, so very few Brits had heard of the game, let alone seen it in action. A charming (but possibly apocryphal) story, related in Roger Fisher and William Ury's negotiating handbook *Getting to Yes*, tells of an American father vacationing with his son in London. The pair head to Hyde Park to play frisbee and their exertions draw a small audience of fascinated Brits who watch the impressive and previously unseen game, enrapt. Eventually, one English onlooker plucks up the courage to approach the father and ask, 'We've been watching the game for fifteen minutes now and we can't work it out – who's winning?'

This anecdote perfectly encapsulates the zero-sum mindset – the predisposition to believe that in every game, for someone to win, someone else has to lose. But what if that weren't always the case? What if we could alter the rules of the game to ensure that everyone wins – a reshaping of the situation which increases the pooled resources on which everyone draws? This is at the heart of resolving common tragedies like overfishing.

During the Grand Banks cod crash in the 1980s, if the Canadian government could have stepped back and viewed the situation objectively, they would have realised that to secure jobs in the industry, the fish stocks had to be allowed to recover. If they had recognised that placing temporary caps on catches was in everyone's future interest, the decline might have been arrested. In the long term, you don't win the fishing game by taking as many fish out of the ocean as you can. You win by keeping the population as high as possible and removing at most the *maximum sustainable yield*. The yield will remain high in perpetuity as the cod stocks remain undepleted. But the costs for cheating in this game have to be big enough, and strictly enforced, so that taking only the allowed catch is in an individual's best interests as well as the collective's.

*

In contrast, there are multiple examples of situations in which even relatively minor tweaks to the rules can completely change the game. Most of us are keen, in theory, to limit our impact on the environment by reducing our use of plastic, particularly single-use plastic. Our intentions, though, are not always borne out by our actions. When plastic shopping bags were given away for free in UK supermarkets, forgetting to bring my old carriers with me carried no penalty. I simply got some new ones at the checkout and added them to the mountainous bags of bags in the cupboard when I got home. And I was not alone in my lazy plastic habit. In 2014, one study suggested that the average UK home had forty plastic bags lying idle around the house. One way or another, despite our best intentions, many of those plastic bags would eventually end up being thrown out, likely ending up in landfill or, worse, in the ocean. Despite everyone already having more bags than they could use, the number of plastic bags given away by supermarkets continued to increase year on year throughout the early 2010s. But one simple change to the game dramatically altered the attitudes and habits of millions of UK shoppers almost overnight.

On 5 October 2015, the UK government brought in a law which meant that large retailers in England had to charge customers 5p for each plastic bag they obtained in the shop. Five pence is not a huge amount of money for most people. Even purchasing ten bags adds only 50p to the price of a shop, but this small incentive to re-use plastic bags (or rather punishment for failing to re-use) made a huge difference to our consumption almost instantaneously.[92] In 2014, the year before the charge was brought in, customers were given over 7.6 billion plastic bags at English supermarkets – around 140 for every adult in England. By 2019, total single-use plastic-bag sales had dropped to just 564 million, fewer than 10 per person per year. In addition, the charge has raised over £180 million for charities across the UK. Everybody wins.

The plastic-bag problem hasn't been completely solved, with many people switching to more durable (but still plastic) 'bags for life' which can be reused multiple times and replaced for free when worn out. This free-replacement scheme means that you are effectively renting your bag from the supermarket and is reminiscent of a deposit model which many countries have employed to reduce marine debris: customers pay a small deposit on the purchase price of a drink which they can reclaim when they return the bottle. The customer doesn't lose out, as long as they return their bottles, the manufacturers save money and energy by reusing the returned bottles, and everyone benefits from the consequent reduction in pollution and natural-resource usage.

If you've ever organised an event, you might learn something from the bottle-deposit scheme, especially if you've suffered with problems of capacity management and low attendance. People booking a place and then not turning up is a particular problem at events for which the tickets are free. Knowing they have not invested in the event financially means that people also feel freer to reclaim the time and effort required to attend by not turning up on the night.

When I held an event at London's Southwark Cathedral to launch my last book, I faced just such a problem. Fortunately, the team helping me to run the event were aware of this potential issue and deliberately overbooked in order to ensure a full house on the night. Despite the fact that the numbers worked out almost perfectly, the uncertainty – wondering whether anyone would turn up at all, or if we would have to turn keen interested parties away – only served to add to my opening-night nerves before the launch.

In hindsight, a simple but practical solution, which has been shown to work for such events, is to charge a small fee for a ticket with the incentive of a refund on the ticket price upon attendance at the event. Interestingly, even in the absence of the refund, a small ticket price has been shown to improve attendance. The theory behind this improvement suggests that without the financial commitment, potential attendees see only the incidental costs of attending – their time and effort – and, without the buy-in of a paid ticket, are prone to save themselves those costs by not attending. A zero-ticket price can have the unintended consequence of lowering overall attendance, despite potentially increasing registrations.[93] This is a classic example of a *boomerang* – a phenomenon we will study in more detail in Chapter 8. People who have paid for a ticket are less willing to sacrifice their initial investment, and more likely to think deliberately and make plans to attend in advance of the event, rather than speculatively assessing whether they can make it on the day. For many events, a small entrance fee will also ensure that the audience who do attend are those who are genuinely interested, rather than people who may have signed up on a whim and will perhaps not be as engaged in the content.

If we think creatively, there are very few scenarios in which we cannot come up with solutions that are mutually beneficial to all the parties concerned – games we cannot change. When it comes to the biggest potential commons tragedy of the lot – climate change – this is something we can't afford not to do. The costs of failure are too great for all concerned. The superpowers of our world have it in their gift to reverse the direction of global temperature rises. Whether they are able to instigate a system which aligns the actions that are in the planet's long-term best interests with those that are to their own short-term benefit still remains to be seen. But by using the framework of game theory to re-analyse the problems the world currently faces, it can be hoped that we will come up with novel ways to change the rules of the game which will future-proof our planet.

6

READING BETWEEN THE LINES

Ninety-one-year-old Briton Margaret Keenan was the first person in the world to receive a dose of the first vaccine approved for use against Covid-19. Understandably, given the impact that the pandemic had had upon the UK's economy up to that point, all eyes were on the vaccination programme. It seemed to be a race between the virus and the vaccine. Three weeks after Margaret received her jab, in the first week of January 2021, the UK was vaccinating 300,000 people a week on average. Everyone wanted to know when, if ever, we were going to have vaccinated enough people that our lives could return to normal.

In the UK, Channel 4 news ran a segment in which they projected how long it would take, at the current rates, for the whole of the adult population of the UK to receive both doses of the Covid-19 vaccine. 'If we continue at that rate [300,000 doses per week], it will take over six years until October 2027.' As it transpired, the UK was able to offer (of course not everyone decided to take up the offer) two doses of the vaccine to the whole adult population by the end of July 2021, six years ahead of the news programme's pessimistic projection.

The UK media were not alone in making these sorts of gloomy

prognostications. Late in December 2020, it became clear that the United States would fail to hit the Trump administration's target of distributing 20 million doses of vaccine by the end of the year. In fact, just over 5 million Americans had received their first dose by 30 December, casting doubt on the feasibility of Operation Warp Speed's target to vaccinate all Americans by June. President-elect Joe Biden was moved to suggest that 'At the pace the vaccination programme is moving now . . . it's gonna take years and not months to vaccinate the American people'. Almost ten years, in fact, according to analysis by NBC news. As it turned out, all adult Americans were offered their vaccine inside ten months, rather than ten years.

So why were these takes on the speed of the vaccine rollout so far off?

The simple answer is that those projections were predicated on a simple mathematical assumption. If you're not careful, though, it's easy to gloss over the fact that a mathematical model is being used in these predictions at all because there are no equations given up front, only phrases like 'at that rate' or 'at the pace'. Hidden behind these phrases is a bias that afflicts so much of our reasoning about the future that most of us don't even realise we are subject to it. You may never have heard of it before, despite using this shortcut all the time. Its name is *linearity bias*.

Linear thinking

The word 'linear' describes a special relationship between two varia-bles – an input and an output. If a relationship is linear, a change in one quantity by a fixed amount will always produce a fixed change in the other quantity. This is a good model for all sorts of real-world relationships. With a fixed exchange rate, £1 might be worth 2NZD (New Zealand dollars), £10 would be worth 20NZD and £100 would be worth 200NZD. As you increase the pounds you want to exchange, the number of dollars you get back increases in proportion. Driving at a fixed speed, it should take me twice as long to reach a destination twice as far away. Time required increases in proportion to the distance travelled at a fixed speed. If I can buy three chocolate bars for £1, then surely, I can buy six chocolate bars for £2. The number of bars I can purchase scales linearly with the money I'm prepared to spend. Linearity assumes there are no three-for-two offers on the table. If the relationship between the two quantities begins with them both being zero – that is to say that for no pounds you get no dollars in return, or in no time I cover no distance – then when you double one quantity, you also double the other. This is known as *direct proportion*.

But linear relationships don't have to be in direct proportion. Take the relationship between the two widely used temperature measures, Fahrenheit and Celsius, for example. To convert from Celsius to Fahrenheit you need to multiply the Celsius temperature by 1.8 and add 32. Normal body temperature is about 37 degrees when measured in Celsius. This corresponds to 98.6 (which is $37 \times 1.8 + 32$) degrees Fahrenheit. But the freezing point of water isn't at the same value for both scales – it's 0 degrees in Celsius, but 32 in Fahrenheit – so we don't have direct proportion. This means that doubling the temperature in Celsius from 5 to 10 degrees doesn't double the temperature in Fahrenheit. Instead, in Fahrenheit the temperature would rise from 41 degrees to 50 degrees. Nevertheless, the linear nature of the relationship means that a fixed change in the temperature measured in one regime always corresponds to a fixed change measured in the other. A rise of 5 degrees Celsius is always a rise of 9 degrees Fahrenheit, no matter what temperature you start from. Given the input in this linear relationship, we can easily visualise the output as in Figure 6-1. These relationships can be represented as straight lines, which is, of course, why we call them linear.

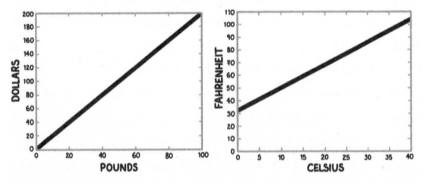

Figure 6-1: The number of dollars you get varies in direct proportion with the number of pounds you exchange (left). The temperature expressed in Fahrenheit varies linearly with the temperature in Celsius, but not in direct proportion – 0 degrees Celsius corresponds to 32 degrees Fahrenheit, not 0.

Perhaps I have laboured the point a little about these linear relationships, especially since linearity is such a familiar idea to us. But herein lies the problem: we are so familiar with the concept of linearity that we impose our linear frame of reference on data we observe in the real world. We assume, because the way something is growing right now appears linear, that the relationship will continue to be linear in the future. This is linearity bias in its simplest form. In some circumstances, it may be the correct assumption. The distance travelled in a fixed amount of time really will scale linearly with the constant speed at which you drive, but many systems do not obey these simple linear relationships. What is worse is that many real-world relationships may look initially as though they are linear, but then veer wildly off their expected course with seemingly little notice.

These are the phenomena that I often refer to as *curveballs*. Just like a curveball in baseball, they initially look like they are headed in one direction, encouraging us to make a prediction about their future trajectory. However, these curveballs have a tendency to swerve off, meaning they don't end up where we expect them to, causing our predictions to miss their targets. This is where we run into problems when making simple extrapolations about the future – when we assume the relationship between two variables is linear without sufficient justification.

Matt Frankel is a certified financial planner based in South Carolina. He makes his living investing in stocks and shares, and by making his financial insights available to the world on his blog. In 2011, he spotted a car company he thought had the potential to make him some serious money. In June 2010, the company had delivered an exciting initial public offering, launching shares at $19 apiece. After a promising first day's trading in which the price rose sharply to $23.89, the next nine months would see the shares stagnate around the $23 mark. This was when Matt took his chance. He invested heavily in the impressive but largely unproven firm and sat back to see what would happen. Sure enough, the share price began to rise steadily. Despite the minor ups and downs experienced by all shares (captured by the solid black line in Figure 6-2), the value rose gradually at a rate of about $4.50 per year on average (dashed grey line in Figure 6-2). Decent, but for a financial planner, not stellar growth.

A little over two years on and Matt's shares had increased in value, but were not performing as well as some of the other investments in his portfolio. He decided it was time to take his money out of the indifferently performing car company (which had still not posted a profit in any quarter up to that point) and put it into something that would give a higher rate of return. In March 2013, Matt sold all his shares in Tesla.

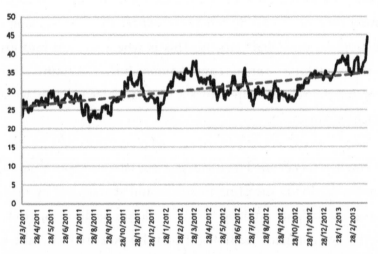

Figure 6-2: Tesla's share price (black) changed approximately linearly over a two-year period while Matt held shares. The dashed grey line is the line of best fit to the share price. The slope indicates a rise in share price of approximately four and a half dollars per year on average.

What Matt didn't know was that this was perhaps the worst time to sell – Tesla's share price was a classic curveball. He had sat through a relatively steady period of slow but sustained growth, the price increasing approximately linearly with time, rising from $23 per share to $40 per share in two years. By 1 October 2013, less than six months after Matt sold his shares, they were worth almost $200 each (as illustrated in Figure 6-3). At the time of writing, each one of Matt's $23 shares would have been worth over $3000.

Looking at the first plot of Tesla's share price up until March 2013, the dashed grey line (describing the average trend once the fluctuations in the data are averaged out) suggests that the share price was increasing linearly with time. But, of course, there is no reason to believe that the share price should continue to grow in the same way, or indeed continue to grow at all. The stock market is a complex, nonlinear system in which hundreds of different variables can impact upon the price of shares. Assuming things will stay the same or that recent trends will continue linearly into the future is a dangerous game, but one which we are preconditioned to play.

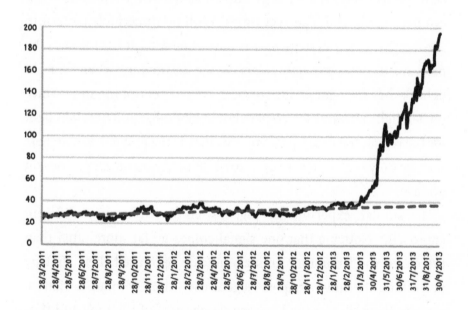

Figure 6-3: Tesla's share price rocketed shortly after Matt sold his shares. The linear projection (dashed grey line), based on two years' slow and steady growth, would have predicted a share price of about $40 by October 2013. In reality, the share price (full black line) had risen to almost $200 by that point.

We have a tendency to base our long-term forecasts on short-term trends. In the 1960s researchers undertook a series of experiments designed to understand human predictive behaviour.[94] In a typical experiment researchers might have asked participants to predict which of a pair of lights would flash next – left or right. Over the course of the experiment the left light was set to flash 70 per cent of the time and the right light the remaining 30 per cent. However, the sequence of flashes was randomly determined and hence unpredictable. After several rounds of the game, most participants ended up guessing left and right with the correct frequencies (70 and 30 per cent, respectively) – a strategy known as *frequency matching* – but not necessarily at the correct times. With no discernible pattern to the sequence, this strategy means that when the left light flashes the participants guess correctly 70 per cent of the time and when the right light flashes they guess this correctly 30 per cent of the time. Since the left light accounts for 70 per cent of all flashes and the right light for the remaining thirty per cent, this means the frequency-matchers average only (0.7 × 70% + 0.3 × 30%) 58 per cent correct guesses on average.

In similar experiments, pigeons employed a very different approach.[95] Noticing that one signal comes up far more often than the other, the laboratory animals quickly optimised their strategy by picking the more frequent signal each time without deviation, earning themselves a food reward 70 per cent of the time and thus far outstripping the humans' success rate. Even after the human participants had been informed that the sequence was randomly generated, and hence unpredictable, they continued to use the sub-optimal frequency-matching strategy in the hope of predicting a non-existent pattern.

The kicker to the experiment was that in the last round, instead of flashing up a predetermined sequence, whichever light the human participants pointed to as their prediction would subsequently light up. In this last round, the human participants continued to use the frequency-matching strategy with the frequencies they had learned previously, but this time, by design, they were 100 per cent successful. When asked for their perceptions of the reasons they had been able to achieve perfect scores in the final round, the participants typically responded that there was a pattern they had finally figured out. They proceeded to describe elaborate and unlikely sequences of left- and right-light flashes that resulted in their choices being correct.

Playing the markets

The flashing-light experiments, just like the random point-scoring experiments in Chapter 1 (which engendered the development of superstitious behaviour) demonstrate that we are predisposed to look for patterns in data. We hope to find trends that will allow us to predict what will happen next, even when there are no such trends to be found. When it comes to investing in the stock market, as Matt Frankel found out to his cost when divesting from Tesla, our propensity to detect and extrapolate short-term movements that may have no persistent cause can lead to bad decision making.

Perhaps the most famous truism about playing the stock market successfully is to 'buy low and sell high'. Of course, if it were that easy, then everyone would do it. Perhaps less well known, but more apposite to explaining the difficulty with predicting the stock market, is the following facetious advice: 'Buy a stock, wait until it goes up and then sell it. If it doesn't go up, then don't have bought it.' Although it becomes obvious in hindsight, the key is knowing when the value of a share or an index is about to bottom out or to peak *before* it happens. No matter how much they believe in their 'strategy', the investors who seem to display this sort of superhuman foresight are typically just riding their luck and find their miraculous feats difficult to repeat. A tempting substitute for us mere mortals is to use short-term trends – to buy shares when they seem to be on their way up and to sell them after an apparent downturn. Appealing as it sounds, this *market-timing* strategy can lead to exactly the opposite of the desired and oft-quoted adage. Instead, we end up losing money by buying high and selling low.

The prices of stocks and shares naturally fluctuate over time. Even share prices that rise in the long-term experience short-term dips. A practical market-timing strategy might be to wait for the price to fall by 5 per cent, say, before selling, and then waiting until it rises 5 per cent before buying again. But how does this compare to a strategy in which you leave your initial investment untouched and ride out the rises and falls? You can compare the impact of the two strategies by looking at a stock which fluctuates, but is expected to rise in the long term, as in the solid black line in Figure 6-4. Selling shares once they have dipped locks in your losses and buying back once you notice them start to rise means you have missed out on the early stage of the upward trend. Generally, by pursuing the short-term market-timing strategy, the falls you negate are smaller than the rises you miss when the share price is climbing overall. Selling when prices start to fall and buying when they start to rise, in response to short-term move-ments, tends to result in long-term losses in comparison to investors who hold their nerve, remembering why they invested in the first

place. Indeed, studies into the performance of active fund managers –
people who will actively invest your money for you for a hefty fee
and who generally purport to know what they are doing – regularly
show that the vast majority of them underperform in comparison to
indexes linked to the stock market as a whole. Doing your research,
investing in apparently undervalued companies and then sticking it
out for the long term is often the best strategy.

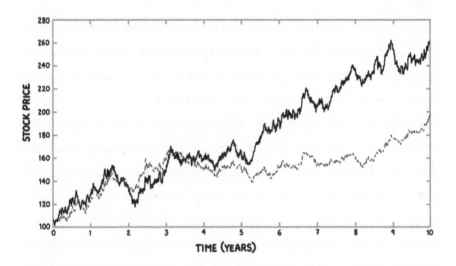

*Figure 6-4: The price of a hypothetical share (solid black line) rises over the period
of ten years. While the market-timing strategy (black dashed line) can outperform
the long-term investment strategy, especially when the stock prices fall (as in year
3), for a stock which shows long-term rises, the market-timing strategy typically
leaves investors out of pocket in the long run.*

Stock market aficionados often talk about the difficulty of outper-
forming an efficient market. The *efficient-market hypothesis* suggests
that share prices reflect all the available information about past,

current and potential future performance of every company in the stock market. The opportunity to spot 'mispriced' stocks, which can be exploited to make a profit, should be limited, then. In some senses, the idea of the efficient market relies on the wisdom of the crowd – that many rational heads can be better than one. However, as we discovered in Chapter 5, individuals acting rationally in their own self-interest don't always achieve the outcome that's best for the group and, as we shall see in Chapter 7, crowds can be remarkably capricious.

Despite being much vaunted for their collective wisdom, crowds are also notorious for their collective madness – their herd mentality. The idea that mass participation in the stock exchange leads to an efficient market in which every company's price reflects its true worth ignores the psychology of the members of the crowd – often ill-informed human beings driven by fear and its opposing emotional state, greed. These two emotions, in response to market behaviour, rather than information about the companies themselves, are widely accepted to be two of the main drivers of volatility in the stock market. Greedy behaviour can lead to investment bubbles, like the dot-com boom of the late 1990s, while fear can lead to them bursting.

*

In an era when the use of the internet saw massive global growth, potential investors watched as share prices of online companies soared. The Nasdaq (a technology-heavy US stock market) rose over 400 per cent between mid-1995 and spring 2000. Investors, many of them first-timers, piled in to buy a stake in companies with no history of posting a profit and sometimes companies that hadn't even made a sale. The more prices rose, the more people wanted a piece of the action, creating a positive-feedback loop (of the sort we will meet again in Chapter 7), leading to absurdly overvalued companies.

Shirley Yanez was one such investor. Already living the high life as the head of her own recruitment company, Shirley decided to invest £90,000 of her capital in shares in the dot-com market. Despite her being relatively late to the party, the value of her shares rose to over £2.5 million over the next eight months. Buoyed by her success, in 1999, Shirley sold her house and her business and ploughed everything she had into the dot-com market. At the peak of the bubble, she was worth over £6.5 million – on paper.

In the spring of 2000, a number of events – including a rise in US interest rates, Japan entering a recession, two of the biggest technology companies (Yahoo and eBay) calling off a merger and media coverage warning of dot-com companies running out of money – led to a crisis of confidence in the overinflated technology companies. Shares plummeted. Many investors who tried to ride out the dips lost huge sums, as company after company went to the wall. From an all-time high in March 2000, by the autumn of 2001 the Nasdaq had fallen over 70 per cent, wiping out most of the extraordinary gains of the late nineties.

In a matter of months, Shirley's investments tanked, making them effectively worthless. Her marriage imploded and she found herself selling her possessions just to pay the rent. In the depths of her despair, Shirley overdosed on painkillers in a desperate bid to free herself from her depression. Fortunately, she survived her attempted suicide and was eventually able to rebuild her life from the ground upwards.

Shirley is not alone in experiencing desperation driven by financial crashes. Suggestions of increased suicide rates in the aftermath of financial downturns have existed since at least the stock-market crash of 1929, which triggered the Great Depression. In 2018, researchers found definitive increases in suicide rates in the general population of developed countries during periods of significant market decline and the following year. For both men and women, suicides were found to increase most significantly in the aftermath of stock-market crashes and banking crises. In the year following the dot-com crash, suicide rates increased by 20 per cent for men and 8 per cent for women, in comparison to what might have been expected had the bubble not burst.[96]

If the litany of burst stock-market bubbles teaches us anything, it's that it's generally not a good idea to look solely at the changes in the price of a stock to decide when to sell. There are, of course, valid reasons for wanting to liquidate investments: perhaps your motive for buying the stock no longer applies or you need to free up the capital. However, fear induced by a short-term movement in share price should not be one of those reasons.

Reciprocal relationships

One of the factors that can catch us off guard when dealing with rises and falls in investments is the asymmetry in the gain needed to recoup a loss. Perhaps surprisingly, even a small percentage fall always needs to be offset by a larger percentage rise in the stock price. Similarly, a seemingly big percentage rise can be wiped out by a smaller percentage fall. At first it seems counterintuitive that, despite the Nasdaq growing over 400 per cent during the dot-com bubble's ascent, a fall of just 70 per cent was enough to erase almost all the gains that were made over the preceding five years, but percentage rises and falls do not combine additively as our linear expectations would lead us to believe.

If we invest £100 in a company and their share price falls by 10 per cent to £90, a 10 per cent rise from that position only takes us back up to £99. The rise required to get us back to £100 is just over 11 per cent. For bigger percentage losses, the commensurate rise required to level up is even larger. For a fall in share price of one quarter, the corresponding rise to recoup the loss is one third. For a fall of 50 per cent, the share price must double – a rise of 100 per cent – to leave us even. When the Nasdaq rose 400 per cent an 80 per cent fall would have been enough to set it back to square one.

The relationship (illustrated for different percentage losses in Figure 6-5) is clearly nonlinear. In this case it is known as a *reciprocal relationship*. The sense in which the descriptor 'reciprocal' should be understood here is in its interpretation as an 'inverse' or 'inverted' relationship. The reciprocal of a number, z, is just one divided by that number, $1/z$ – what in mathematics we would call the *multiplicative inverse*. For example, the reciprocal of two is a half, $1/2$. When two reciprocal numbers are multiplied together, they return us to one – the whole. To recoup a loss of half $(1/2)$ in a share price (a 50 per cent decrease) the share price must double ($\times 2$ – a 100 per cent increase) – a nonlinear, reciprocal relationship.

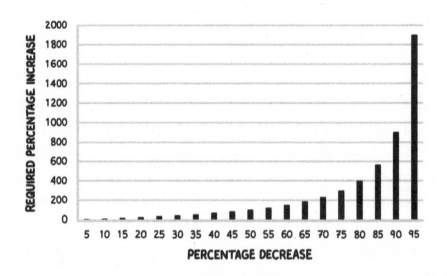

Figure 6-5: The reciprocal, nonlinear relationship between a percentage loss and the pecentage gain required to recoup it.

Exploiting our propensity to think linearly in this context is one way in which companies offering financial products can cook the books. By showing a percentage return averaged over a number of years, they can make their performance look better than it actually is. For example, a fund which gains 50 per cent in one year and loses 50 per cent the next year doesn't break even at the end of two years. The 50 per cent gain and the 50 per cent loss don't simply add together and cancel each other out. Instead, the relative gain and loss must be multiplied together: 50 per cent of 150 per cent isn't 100 per cent – it's 75 per cent, corresponding to a 25 per cent loss over the two years.

These sorts of tricks are also exploited by organisations or individuals wishing to push a particular agenda when the facts don't support their viewpoint. Writing in *The Spectator* under the headline 'The

Brexit bounce is underway' – referring to the supposed positive impact on the UK economy of leaving the EU – Wolfgang Münchau claimed: '... UK exports have made a near-complete recovery. They were up 46.6 per cent in February after falling by 42 per cent in January.'

Quite apart from the fact that a return to the status quo wouldn't constitute what most people would consider a 'bounce', Münchau's argument exploits the nonlinear relationship between the expected rise required to offset a given fall. The 46.6 per cent rise in exports sounds bigger than the 42 per cent fall. In fact, according to the Office for National Statistics, exports fell 42 per cent from £13.6 billion in December 2020 down to £7.9 billion in January 2021. They then rose 46 per cent from that lower baseline, to £11.6 billion in February 2021, constituting an overall fall of 15 per cent from the December 2020 figure. In economic terms, a 15 per cent fall in trade is absolutely enormous. Its apparent size is only diminished in comparison to the extraordinarily unprecedented size of the drop in January. Certainly, there aren't many economists who would call being 15 per cent down overall 'a near-complete recovery'.

Nonlinear, reciprocal relationships occur in many other everyday contexts where they have the potential to confound our typically linear thinking. Imagine, for example, that you've just started a job as the IT manager at a large company whose offices span several different sites. Each of the sites you manage typically downloads 1000 gigabytes of data per day. Your aim is to minimise the time your colleagues have to wait for their downloads. Just before you started, the previous IT manager upgraded the download capacity of half the sites to 200 gigabytes per hour (GBph), while the remaining ones were left with their original 100 GBph speed. Your boss has given you just enough money to upgrade half of the sites again, specifying the following two options:

A. Upgrade all the 200 GBph connections to 500 GBph, or
B. upgrade all the 100 GBph connections to 200 GBph.

Which is the better strategy to reduce the overall time your colleagues spend waiting for their downloads?

If you're anything like me, you'll probably have plumped intuitively for option A. Improving the bandwidth at half of your sites by 300 GBph (from 200 GBph to 500 GBph under option A) seems like a better plan than improving download speeds for the other half of the sites by just 100 GBph (from 100 GBph to 200 GBph under option B). On top of this, the relative improvement in download speed of 2.5 (500/200) for option A compared to 2 (200/100) for option B makes option A seem like a no-brainer. In fact, B is the better investment – and by a big margin.

This comes as a surprise to most people because without thinking about it carefully we assume that the download time is a simple linear function of download speed. If we increase the gigabytes per hour by a constant amount, we expect the download time to decrease by a constant amount (as in the left-hand panel of Figure 6-6), but this is far from the truth. In fact, the download time is a nonlinear, reciprocal function of download speed. For a given increase in download speed, the corresponding decrease in download time depends heavily on what the download speed was to begin with (as in the right-hand panel of Figure 6-6).

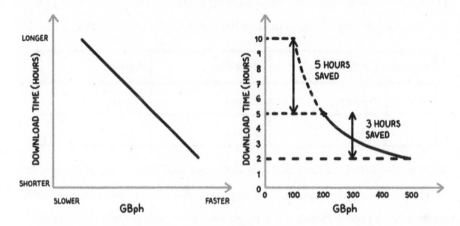

Figure 6-6: The linear variation we imagine that download time should exhibit with download speed (left panel) and the way in which download time actually varies, as the reciprocal of the download speed (right panel). A fixed change in download speed produces a different change in download time, depending on the original download speed.

When your predecessor upgraded half the sites' capacities from 100 GBph to 200 GBph, the new faster connections at these sites reduced their download time by half: from ten hours a day to five – a saving of five hours a day. By choosing option B, you stand to make the same improvement for the sites which currently have the 100 GBph connection, as you can see in the second row of Table 2. Since the total download time at the sites with the 200 GBph connections is only five hours to begin with, to produce the same savings by upgrading these sites, downloads would need to take up no time at all – equivalent to having an infinite download speed. As can be seen in the top row of Table 2, the reduction in time achieved by replacing 200 GBph connections with 500 GBph connections is just three hours. No matter how fast the upgraded connections at the 200 GBph sites – 1000 GBph or even 1,000,000 GBph – it's always going to make more sense to replace the 100 GBph sites' connections with 200 GBph connections.

OPTION	CURRENT	UPGRADED	SAVING
A	5 hours	2 hours	3 hours
B	10 hours	5 hours	5 hours

Table 2: The download time (hours to download 1000 GB) of one of the sites to be upgraded under each option. Replacing 100 GBph connections with 200 GBph connections (option B) yields much bigger time savings than replacing 200 GBph connections with 500 GBph connections (option A).

My wife and my daughter are both redheads. As a family, we are acutely aware of the risk of sunburn and skin cancers posed by unprotected exposure to the sun's rays. Of the two types of ultraviolet radiation that reach the earth's surface – UVA and UVB – UVB plays the most significant role in causing both conditions. As a consequence of years spent sitting largely unprotected in the sun, on holiday or in the garden (on even the few sunny days a year my native Manchester has to offer), my dad regularly has to have basal-cell carcinomas removed from his skin before they become invasive. When I go away with my own family, as a matter of course, we pack the factor-50 sunblock, even if we are holidaying in the UK.

The sun-protection factor (SPF) numbering on sunscreen can be confusing. The higher the number, the more damaging UVB radiation is blocked, but the relationship between the number on the bottle is not directly proportional to the amount of radiation screened out. Factor 50, for example, is not twice as effective as factor 25 at blocking UVB radiation. Factor 30 does not block three times as much UVB radiation as factor 10. When applied correctly, factor 10 blocks out 90 per cent of all UVB radiation. Factor 30 blocks out just over 97 per cent and factor 50 blocks out 98 per cent. The higher you go, the smaller the level of increased protection you are afforded – another case of diminishing returns caused by a reciprocal relationship. So the increase in SPF from 10 to 30 gains you over 7 per cent more protection; however, the increase by the same numerical margin from 30 to 50 gains you less than 1 per cent extra sun-screening effectiveness. Factor 30 is usually the baseline SPF recommended by dermatologists. Lower than that and the degree of protection afforded starts to drop off quickly.

The way SPFs are often explained is by talking about the increase in exposure times different factors allow. If your skin could sustain ten minutes of exposure before burning without any protection, then the idea is that SPF 10 would extend that time by a factor of 10 to 100 minutes, while SPF 50 would extend it to 500 minutes. The underlying maths is that you can find the total UVB radiation exposure by multiplying the exposure time and the intensity of radiation experienced. When you apply SPF 50, the duration of time you can theoretically spend in the sun without getting burned increases by a factor of 50 (hence it is called a sun-protection *factor*). If the total exposure is to be the same, to compensate for this increased time, the intensity of radiation must decrease by the same factor – 50. The relationship isn't linear. Factor 10 lets only 1/10 of the radiation through, blocking 9/10 or 90 per cent. Similarly, factor 50 lets only 1/50 (or 2 per cent) of the UVB radiation through, which is where the figure of 98 per cent screening effectiveness comes from for factor 50. Just like the relationship between the time taken and the speed you travel at, the relationship between radiation intensity and sun-protection factor is reciprocal.

Mathematics aside, it's sensible to exercise caution in the sun. The

SPF only refers to protection against UVB rays which cause most skin cancers and sunburn. It does not provide protection against the deeper-penetrating UVA rays, which are largely responsible for premature skin ageing, but can also cause some skin cancers and contribute to sunburn. Most dermatologists would recommend reapplying sunscreen every two hours, since protection can diminish over time as it breaks down, dries out or is rubbed off your skin. By talking about the linear relationship between SPF and extended duration of exposure (the idea that factor 10 allows you to stay out ten times as long), rather than screening efficiency, the traditional explanation of SPF is misleading – precisely because the effectiveness diminishes over time as the sunscreen wears off. We gain a false sense of security about how long we can stay out in the sun safely.

Most of us don't appreciate these reciprocal discrepancies when making forward-looking decisions on the benefits of faster internet connections, vehicle-fuel efficiency or even the sun-protection factors of sunscreen. As a consequence, we leave ourselves vulnerable to linear price hikes for attenuating nonlinear benefits.

Knowledge is power

At least part of the reason why we assume that quantities vary linearly is because linear relationships are deceptively familiar. We learn the rules of straight lines very early on as children. The shortest path between two points is the straight line that connects them. It's easy to tell just by looking at something if it is a straight line, and just as easy to describe its shape accurately to someone else. The same is not true for an object which is curved. The problems we solve in our early maths lessons are linear. If Jane pays £5 for ten grapefruits, how many grapefruits does she get for £50? In this idealised linear mathematical world, there are no discounts and nobody bats an eyelid when you buy 100 grapefruits.

In fact, our tendency towards a linear frame of mind goes way beyond our early-years experiences with these straight-line relationships. It is far more deeply ingrained than that. Believe it or not, when thinking about relationships between two quantities (an input and an output), we have preconceptions about the size of the output we expect for a given input. We can reveal these expectations using an experimental technique called *iterated function learning*.

A mathematical function can be thought of as a simple drawing machine which plots a single output, y, for each input, x. The simplest member of this family of drawing machines is the constant function which just gives you the same output, irrespective of the input. The constant function draws a horizontal line across the page (as in the first panel of Figure 6-7). A constant function might be used to describe the relationship between the size and value of items in a pound store. Whatever the input (whatever the size of the object you want to buy) the output (the price) is always the same – £1.

The next simplest is probably the function whose output depends linearly on the input. This function draws a straight diagonal line across the plot. If the linear relationship is in direct proportion, then the line extends from the bottom left of the plot to the top right, going through the point zero-zero (as in the second panel of Figure 6-7). This sort of function might be used to help you calculate how much fuel you need for a long journey. If you know the miles per gallon expected from your car, then given journey length as an input, a linear function can be used to calculate the output of how much fuel you will need.

The *quadratic function*, for which the output is the square of the input (giving the increasingly steep curved line in the third panel of Figure 6-7), is also useful for drivers. Just such a relationship describes how the braking distance (the distance travelled once the brakes are applied with constant force until the vehicle comes to a complete stop) increases with speed. It's especially important for new drivers to understand that doubling their speed doesn't double the braking distance but instead quadruples it.

Another more complicated function is the *sine function* whose output see-saws in size as the input increases to plot an oscillating curve across the page (as in the fourth panel of Figure 6-7). The sine function might be used to describe roughly how the number of hours of daylight changes across the year, increasing from the spring equinox up to a maximum at the summer solstice and then back down through the autumn equinox and dropping further to a minimum at the winter solstice, before rising smoothly back to where it started at the next spring equinox.

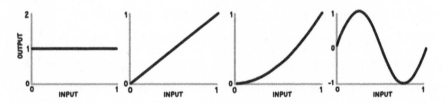

Figure 6-7: Different functions specify different outputs for a given input, drawing different curves across the plot. From left to right: the constant function, a linear function (representing direct proportion), a quadratic function and a sine function.

It's no easy task to understand what people's inherent biases about mathematical functions are. Iterated function learning experiments are like controlled games of Chinese whispers played in the lab. One participant has outputs for a number of randomly selected input points from a given stimulus function sequentially flashed up at them on a screen. When the sequence is finished, they try to reproduce the outputs for a regular set of input points. Their effort is used to generate a sequence for the next participant who, again, attempts to reproduce what they have been shown for the next participant and so on. The process is iterated until the message stays roughly the same from participant to participant.

Abstractly, it can be shown that the final answer converged upon in these iterated experiments is reflective of the participants' prior beliefs or biases of what the stimulus should be. When the initial stimulus is a mathematical function, the experiments should converge on people's preconceived idea (explicitly acknowledged or not) of what the function will look like, even before they have received any information.

Irrespective of the shape of the initial-stimulus function, almost universally, after nine (or often fewer) iterations of this mathematical Chinese whispers experiment, the function that is converged upon is the straight line indicating direct proportion.[97] You can see what this progression might look like for four different examples of initial stimulus functions in the four rows of Figure 6-8. For each different initial stimulus function, a relationship which is a good approximation to a straight line is converged upon. The fact that, under experimental conditions, this linear function is almost always the end result is reflective of people's inherent preference, or prior expectation, that a relationship between two variables will be linear.

We use this innate mathematical shortcut to help us extrapolate into the future (as we saw with estimates of future vaccine delivery speed) or to fill in the gaps where there are missing data. Sometimes this linear model is correct and sometimes, as we have seen already, it isn't.

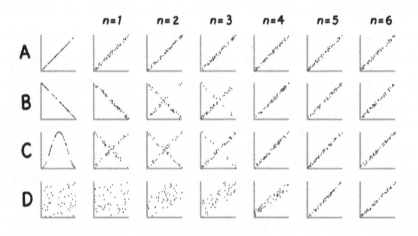

Figure 6-8: An example of what the output of iterated function learning experiments might look like. Irrespective of the initial function (A) linear function with a positive slope; B) linear function with a negative slope; C) sine wave; D) a random set of points), a directly proportional linear relationship tends to be converged upon. Each column after the first shows the output that might be produced by each generation of learners based on the input they were shown from the column to its left. The left-most panels show examples of the initial input function that might be shown to the first participant.

*

The unconscious predisposition towards linear thinking is clearly deeply ingrained in the adults who undertook the function-learning study. Other investigations into the origins of this bias have shown that our propensity to assume linearity is present long before we leave school.[98]

These studies pose students questions in which linearity is not the right tool to use in order to see how they respond. At one end of the scale, these so-called *pseudo-linearity problems* might take the form: 'Laura is a sprinter. Her best time to run 100 metres is thirteen seconds; how long will it take her to run 1 kilometre?'

It is not possible to ascertain the correct answer from the information in the problem. However, most students still reach for the linear solution, without any concern for the unrealistic nature of their underlying assumptions. They scale up the time to run 100 metres by a factor of 10, to account for the distance being ten times longer, giving a time of 130 seconds to run a kilometre. Clearly, this can only ever be an underestimate of the true answer, since it neglects to take into account the fact that no athlete can sustain their best 100-metre pace over the course of a kilometre. Indeed, the linear answer would see Laura breaking the world-record time for the kilometre of two minutes and eleven seconds.

At the other end of the scale are the problems for which the true answer is available.[99] However, it is only when stepping back from the problem that it becomes manifestly clear that the gut instinct to assume a linear proportional relationship is incorrect: 'It takes three towels three hours to dry on the line; how long does it take nine towels to dry?'

Many students reach for the familiarity of proportionality, tripling the drying time because the number of towels is tripled when, in reality, it shouldn't take longer for nine towels to dry in parallel than three.

Part of the linearity bias results from the simple intuitiveness of linear relationships. Despite students undergoing formal learning that suggests there are situations in which linear relationships are not appropriate, for most pupils, these sorts of intuitive ideas have an obvious and almost coercive nature, leading to an unquestioning confidence in their use. In some studies, linearity bias was found to be so seductive that even when confronted with the correct solution by the researchers carrying out the study, many students were reluctant to abandon their original answers.[100]

It seems that the most important explanation for our over-reliance on linearity comes from the mathematics classroom itself. For much of our school mathematics careers, linearity is drilled into us, so that we expect to see it everywhere. Although linearity is an important concept, this 'everything-is-linear' approach fosters the *illusion of linearity* in which students believe a linear model is appropriate for every problem.[101] We are taught to be confident that if it takes twenty minutes to walk a mile, it should take forty minutes to walk 2 miles. If not, we are led to believe that something fishy is going on.

However, this reinforcement of linearity does leave us open to unthinkingly applying our favourite rule, even when it's not appropriate. A compounding factor is the lack of acknowledgment in maths classes that the real world is usually not as simple as a maths problem. For example, to answer the question 'If it costs £1 to send a letter 80 miles from Manchester to Coventry, how much does it cost to send a letter 160 miles from Manchester to London?' relies not on the linearity principle, but on real-world considerations that mean it usually costs the same to send a letter from anywhere to anywhere within the same country.

*

Another type of question used to probe our over-reliance on linearity is as follows: 'Farmer Jones takes one hour to mow a square field with sides of length 100 metres. How long would it take her to mow a square field with sides of length 300 metres?'

The overwhelmingly tempting answer to give is three hours. Just scale the time up by the ratio of the length of the sides of the fields. Indeed, over 90 per cent of thirteen-to-fourteen-year-olds and over 80 per cent of fifteen-to-sixteen-year-olds fell prey to this deceptive linear logic.[102] In fact, it makes more sense for the time taken to mow the field to be proportional to its area, which scales not with its length, but with the *square* of its length. A more correct answer should be nine hours – because a field with sides three times as long has nine (3^2) times the area.

Far from being an abstract maths problem, this sort of area scaling misunderstanding crops up all over the place in real life and we can exploit it to our advantage. Imagine you're hoping to get takeout pizza with three friends. Perhaps you decide to get four 8-inch-diameter pizzas for £10 each. If you can bear to have the same toppings as your friends, though, you might be better off getting one 16-inch pizza for £20. The diameter of the 16-inch pizza is twice as big as the 8-inch versions and the price is also doubled. However, while the price scales linearly with the diameter of the pizza, the area scales as its square (according to the formula $\pi \cdot r^2$, where r – the radius – is half the diameter of the pizza). This means that for only twice the price you get four times as much pizza. The area of the pizza grows by the square of the increase in the diameter, a factor of 4, while the price only doubles in line with the increase in diameter. And there's more good news for people who don't like crusts. Since the length of the crusty circumference of the pizza scales linearly with the diameter (according to the formula $2 \cdot \pi \cdot r$), assuming the crust is of constant width, you also get a better toppings-to-crust ratio with a larger pizza. It's a win–win.

In 2014, Quoctrung Bui, then working as a graphics editor for National Public Radio in the US, decided to go big on the pizza-value-for-money question. He gathered together the prices of 74,476 pizzas from 3,678 different pizza parlours around the US. He found that the price of pizza really does seem to vary linearly with the diameter, as displayed in Figure 6-9, working out at just over a dollar per inch of diameter for pizzas above 8 inches wide.

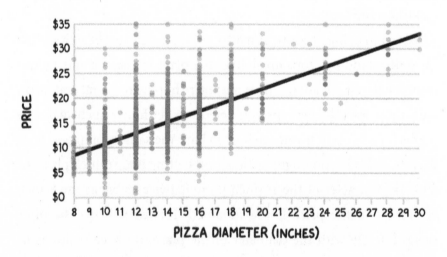

Figure 6-9: The price of a pizza scales approximately linearly (or perhaps I am just preconditioned to impose this relationship!) with its diameter. The grey dots show the prices of different pizzas with the given diameter and the black line is a line of best fit through the points. On average, each inch added to the diameter adds a little over a dollar to the price.

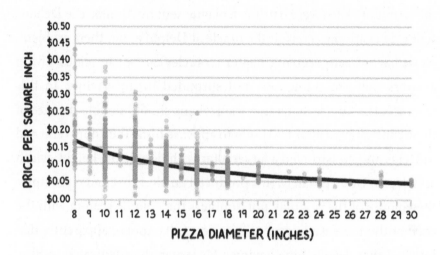

Figure 6-10: The price per square inch decreases as the diameter increases because the area scales with the square of the diameter, wheras the price scales roughly linearly with the diameter.

Naturally, then, this means that the price per square inch decreases because, as the pizzas get bigger (as shown in Figure 6-10), the area increases faster than the price. In terms of value for money, when it comes to pizzas, it always pays to go large.

I should have known that there would be some interesting mathematics underlying pizzas. After all, using the formula for the volume of a cylinder, $\pi \cdot r^2 \cdot h$, where r is the radius and h is the height, we can calculate the volume of a pizza with radius z and depth a, using the formula $pi \cdot z \cdot z \cdot a$.

These sometimes-counterintuitive scaling relationships are nothing new. Indeed, legend has it that the ancient Greeks – famed for their mathematical prowess – fell victim to a mistake caused by misunderstanding a similar property. In one retelling of the story, the people of the island of Delos were battling a plague sent by Apollo. The Delians sent a delegation to consult the oracle at Delphi about their problem. As we have already seen in Chapter 1, oracles can be capricious, giving answers that are ambiguous and shrouded in uncertainty. In this instance, the Oracle suggested the Delians should double the size of their cube-shaped altar to Apollo to appease him. They duly returned to Delos and constructed a larger altar with double the height, width and depth. Unfortunately, this produced an altar that had eight times the volume of the original, not double – the volume scaling with the cube of the increase in each dimension (2^3). Apollo, apparently dissatisfied that despite now having a far larger altar dedicated to him, the Delians had not solved his problem accurately, allowed the plague to continue unabated.

In another retelling attributed to Plutarch, there was no plague, just political tension in Delos. In this version, the problem of *doubling the cube* – sometimes known as the *Delian problem* – was solved correctly with the help of three mathematicians recommended by Plato: Eudoxus, Archytas and Menaechmus. The Oracle's lesson was supposedly to teach the Delians to concentrate their energies on geometry, so as to distract them from their political machinations – but this doesn't make such a good story.

The bigger they are . . .

Perhaps one of the most important nonlinear relationships we encounter regularly is the *square-cube law*. When an object undergoes an increase in size by a fixed-scale factor across each of its dimensions, its new surface area will be increased by the square of the scaling factor, while the volume will be increased by its cube. If a box, like the Delian altar, has its sides doubled, its area increases by a factor of 4 (2^2), while its volume increases by a factor of 8 (2^3).

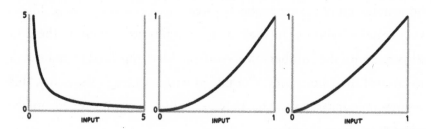

Figure 6-11: Some of the nonlinear relationships which confound us in everyday life, from left to right: the reciprocal relationship which describes how shielding from UVB radiation changes with sun-protection factor; the quadratic relationship which describes how the amount of pizza you get increases with radius; the square-cube relationship which places restrictions on how much we can scale things up without breaking them.

This simple, nonlinear scaling rule (shown in the right-hand panel of Figure 6-11) places some interesting constraints on the biology we would expect to evolve on earth. Perhaps the most important biological square-cube law is that weight depends strongly on a creature's volume, whereas strength depends largely on the cross-sectional area of bones and muscles. As an animal gets proportionally bigger, its volume grows faster than the area of its bones and muscles, making it more difficult for bigger animals to support their own weight. It's no coincidence that the biggest animals that have ever lived are sea creatures. The buoyancy provided by the water means their skeletons have to support less of their weight, so they can grow larger without experiencing the same difficulties that land animals of the same size would suffer. In fact, bony skeletons are superfluous for many sea creatures. Even some of the largest animals in the ocean have no bones to support their bodies. Sharks, for example, take advantage of the buoyant effect of the water in which they live by eschewing bone in favour of more flexible, but weaker, cartilage. Cartilage allows sharks to be lighter than their bony counterparts, enabling them to dispense with the cumbersome swim bladder employed by many fish to control their buoyancy. Combined with cartilage's flexibility, this reduction in mass makes sharks hugely agile – a significant advantage when it comes to hunting.

One of the most infamous and pernicious misapplications of a square-cube law is the *body mass index* (BMI) measurement. This is calculated as the ratio of an individual's mass (proportional to the volume) divided by their height squared. The application of BMI as a diagnostic tool suggests that there is a healthy range for this ratio to fall into. If your BMI is either too low or too high, you can be classified as unhealthy. But why should your mass, which is roughly proportional to your volume (a length cubed) scale with the square of your height (a length squared)? One human, all of whose body dimensions are double another's, might expect to have eight times the volume and hence, all things being equal, eight times the mass. But the square of their height will only be four times as big, so the bigger person might expect their BMI to be twice as large as the smaller person's, despite being no fatter or thinner, relative to their size. This suggests we should be dividing their mass by the cube of their height instead of the square.

However, this argument is not quite correct either. Taller people are not simply scaled up versions of shorter people but tend to have narrower frames relative to their height. Indeed, applied mathematician Nick Trefethen of the University of Oxford suggests that instead of dividing mass by height to the power of two (height squared) or height to the power of three (height cubed), we should be dividing by something in between – height to the power of 2.5. He suggests this would be a solution to the problem caused by BMI that 'millions of short people think they are thinner than they are, and millions of tall people think they are fatter'.

Unsurprisingly, given how its value is expected to change with height, even for people with the same percentage body fat, BMI is not a good indicator of cardio-metabolic health, misclassifying many otherwise healthy individuals as under- or overweight and vice versa. Robert Wadlow, at 2.72 metres tall and weighing almost 200kg, is the tallest human ever recorded. Despite being slim, his BMI at his tallest was 27 kg/m^2, putting him firmly in the 'overweight' category by today's standards. For context, back in the 1940s, when Wadlow died, the average global BMI was around 20 kg/m^2.

Although it doesn't work well as an indicator of human health when used to calculate BMI, something approximating the square-cube law does capture the trade-off between mass and strength quite nicely. In humans, growing significantly taller than average leads to a litany of health-related problems for precisely this reason. Robert Wadlow suffered numerous size-related health issues, including loss of much of the sensation in his legs and feet. He also needed leg braces and a walking stick to support his mobility. Indeed, a poorly fitting leg brace eventually caused Wadlow's death; the brace caused chafing and led to the formation of a blister on his ankle which went unnoticed by Wadlow, due to his numbness. The blister became infected, eventually leading to sepsis. He was twenty-two when he died.

Only two of the top twenty tallest humans ever to have lived survived past the age of fifty and none made it to sixty. Other holders of the world's-tallest-man title have reportedly suffered from scoliosis (spinal curvature) and many different types of back and joint pain, as well as being especially susceptible to the impacts of accidents. The world's second-tallest living man, Morteza Mehrzadselakjani, suffered a serious pelvis injury in a bike accident when he was just fifteen. As a result, one of his legs stopped growing and his left leg is now 6 inches longer than the right.

The bigger someone is, the more damage they can do to themselves through even a relatively innocuous accident. Despite toddlers falling over and bashing themselves regularly, the injuries they sustain are rarely serious. Their relatively thick bones in comparison to their mass means they rarely build up enough energy, even at top speed, to do themselves much damage. Because of their increased mass (compounded by the fact that they are falling from a greater height and that their reactions may be slower, as nerve impulses have further to travel), adults falling over will impact the ground with a much larger force. The nonlinear relationship between mass and bone strength means that although their bones are thicker than a toddler's in absolute terms, they may not be relatively thick enough to compensate for the larger impact caused by their increased mass. For the same reasons, taller people have been found to suffer more fall-related injuries – like hip fractures – than shorter people.

. . . the harder they fall

On a dark and stormy evening in January of 1962, thirty-two-year-old Frane Selak boarded a train from Sarajevo to Dubrovnik. As the train raced through a canyon, a fault on the line caused it to derail. Selak's carriage was pitched into the icy river that ran alongside the tracks and, as it plunged into the freezing water, Selak blacked out. The next thing he remembers was waking up with a broken arm and mild hypothermia. A stranger had saved Selak's life, pulling him to safety, while seventeen other passengers drowned. If Selak's stories are to be believed, this would be just the first of seven brushes with death he would experience over the coming decades. Surviving bus crashes, car fires and explosions led to him being dubbed 'the world's unluckiest man', although the fact he lived to tell the tale each time suggests that the opposite of that moniker would be just as appropriate. However, these later close shaves pale into insignificance in comparison to his most spectacular escape.

In 1963, a little over a year after his miraculous river escape, Selak learned that his mother was gravely ill. Despite never having flown before, he immediately decided to catch the next plane from his home in Zagreb to Rijeka to be at her side. When he arrived at the airport, the first available flight was already fully booked and, rather than face the prospect of sitting waiting in an airport lounge while his mother lay dying, he managed to convince the airline's staff to let him squeeze in with the cabin crew in the back of the plane. For most of the journey, the flight was uneventful – until both engines suddenly and simultaneously failed. As the plane began to lose altitude, the cabin depressurised causing one of the rear doors to malfunction, sucking both Selak and a nearby air stewardess out into the atmosphere. Selak told the *Daily Telegraph* in 2003: 'One minute we were drinking tea and the next, the door was ripped open, and she was sucked into mid-air followed shortly by me'. The stricken aircraft was forced to make an emergency crash landing killing seventeen passengers and both pilots. The stewardess who was sucked out first also perished in the accident, but Selak apparently beat the odds again. After free-falling, he recalls landing in a haystack which broke his fall. By falling through the haystack, his deceleration from terminal velocity to zero was spread over a longer period of time than it would have been had he hit the ground directly, meaning he experienced nowhere near the force that would have accompanied a ground-impact, sparing him almost certain death.

Doubt has since been cast on the veracity of Selak's claims, not least because there is no record of a plane crash anywhere in Croatia in 1963. However, if Chapter 2 and the law of truly large numbers have taught us anything, it's that given enough opportunities, series of incredibly unlikely events – like surviving multiple accidents – really can happen.

Whether his extraordinary free-fall survival story is true or not, better-evidenced tales of individuals who have fallen from a great height without a parachute and survived, although rare, are not completely absent. For example, on Christmas Eve 1971, the flight Juliane Koepcke and her mother were taking from Lima to Pucallpa was struck by lightning. The plane broke up in mid-air, jettisoning passengers, crew and luggage over the Peruvian jungle. Still strapped to her seat, the teenage Juliane fell over 3000 metres to the canopy below. Amazingly, she suffered only cuts and bruises and a broken collar bone as a result of the fall. Despite her injuries, she managed to endure eleven days in the jungle, following streams and rivers, until she was eventually discovered by a group of fishermen and returned to civilisation to be reunited with her father. Although she didn't know it at the time, her mother also survived the fall, but died of her injuries a few days later. Indeed, it is thought that up to fourteen other passengers survived the initial fall but died before they could be reached by rescue parties. Despite these dramatic and high-profile stories of survival against the odds, living to tell the tale after a fall from a plane without a parachute is incredibly rare.

Of course, skydivers, the daredevils who voluntarily throw themselves out of planes, make useful case studies for free-falling. In a vacuum, all objects are accelerated at the same rate due to gravity. Given the same starting position and time, a dropped bowling ball will hit the ground at the same time as a dropped feather in a vacuum. But skydivers are not falling in a vacuum; they are instead heavily reliant on the effects of air resistance. The faster they fall, the larger the resistance to their motion provided by the air, until the point at which the upward force due to air resistance balances the downward force due to gravity. At this point, the diver has reached *terminal velocity*. The upward force due to air resistance depends on the cross-section of the jumper. In a tuck position, their surface area will be smaller and their terminal velocity faster. Conversely, in a spread-eagle position, their terminal velocity will be slower, due to the stronger counterbalancing force provided by air resistance.

The downward force will increase with a jumper's mass whereas the upward force, counteracting gravity, will increase with their surface area. The square-cube law says that the increase in a larger person's surface area is not enough to offset the pull of gravity due to their extra mass – proportional to their volume – so they will end up at a faster terminal velocity. More succinctly: the bigger they are, the harder they fall. One facetious and entirely impracticable piece of advice given to skydivers contemplating a failed parachute opening is 'be small'. The smaller you are, the greater your surface-area-to-volume ratio will typically be and the slower your terminal velocity.

'Land on something soft' – like Frane Selak's haystack – is another glib recommendation.

We know from experience that an insect falling, even from many hundreds of times its own height, will effectively sustain no damage when it hits the floor. Insects' large surface-area-to-volume ratios mean they enjoy relatively low terminal velocities. This is true even for many small mammals. Mice, for example, are typically able to walk away from falls from any height, but as we saw in Chapter 3, when we looked at cats' survival rates from high-rise falls, for larger mammals there is a limit. As J. B. S. Haldane wrote in 1928 in his monograph *On Being the Right Size:* 'You can drop a mouse down a thousand-yard mine shaft and, on arriving at the bottom, it gets a slight shock and walks away, provided that the ground is fairly soft. A rat is killed, a man is broken, a horse splashes.'

The square-cube law has also proved a stumbling block in the past for engineers, and none more so than Nazi architects and designers during the Second World War. Despite making rapid gains following the invasion of the Soviet Union in June 1941, the German army failed to end hostilities on the Eastern Front conclusively over the following months. As the conflict dragged on into 1942, Soviet tanks reached the front lines and began to play a decisive role in winning battles against the Germans. It became increasingly clear that, to turn the tide back in their favour, the German Army would need to up its game by building a tank that was bigger and more heavily armoured than anything that had gone before.

It was in response to this challenge that the audacious idea of the ironically named 'Maus' (Mouse) sprang forth. The mega-tank was designed to be 10.2 metres long, 3.63 metres high and 3.71 metres wide. When completed, it would weigh 188 tonnes. For perspective, the heaviest German tank in operation up to that point, Tiger I – at 6.3 metres long, 3.56 metres wide and 3 metres tall – was roughly a third of the weight, at just 57 tonnes. The armour for the Maus was intended to be 200 millimetres thick in places, compared to a maximum of 120 millimetres for the Tiger. The Maus was on a completely different scale to any other tank.

Almost from the earliest prototyping stages there were problems. The hugely increased weight of the outer armour meant that traditional engines would not be powerful enough to get the Maus moving. In the end, over half of the internal volume of the behemoth was given over to the engine, further increasing the overall weight. Despite the huge inconvenience of having to sacrifice so much of the internal space, the Maus' top speed was still limited to a lumbering 12 mph, less than half the Tiger's sprightly 28 mph. While the Maus' weight had increased over three-fold in comparison to the Tiger, the cross-sectional area of the base increased by around half that amount. To ensure the tank did not routinely sink into the ground, the tracks were made 1.1 metres wide – the two tracks spanning well over half of the width of the vehicle in total. Even then, the Maus would occasionally sink into ground that was not extremely firm or tear up the roads on which it was driven. Designer, Ferdinand Porsche, also struggled to build suspension robust enough for it to support its own weight. Being too heavy to cross bridges, the Maus had to be specially modified to be able to ford rivers, with a snorkel adaptation to provide air to the operators if it became completely submerged.

All these design challenges contributed to delays in development. By the time the first two prototypes of the super-tank had eventually been fine-tuned for operational duty, it was already the middle of 1944. By that point, the Axis forces had begun to falter and were conceding ground on all sides. Not long after their completion, when it became clear that the Soviets would eventually overrun Germany on the Eastern Front, the two Maus prototypes were blown up in an attempt to protect the Nazis' military secrets. Neither prototype ever saw service in an actual battle. Ironically, the tanks proved so hard to destroy that the Soviets were able to reconstruct them. One of them is, to this day, on display in a museum in Kubinka, near Moscow.

Hitler's imprudent obsession with oversized construction was perhaps never more evident than in the plans for the reconstruction of Berlin that he had his chief architect, Albert Speer, draw up. Renamed Germania, the super-sized city would serve as the capital and focal point of the Third Reich – the Greater Germanic Empire. The whole redesign was oriented around the 3-mile-long 'Avenue of Splendours', running north to south through the city. At the northernmost end of the avenue, on the north side of the enormous 'Grand Plaza', was to be a building designed by Hitler himself – the 'Volkshalle' or 'People's Hall'. Inspired loosely by Hitler's admiration for Hadrian's Pantheon in Rome, the huge domed assembly hall was planned to host gatherings of over 180,000 people. Although, in reality, construction never even started, the building was depicted, in the fictional world of Robert Harris' *Fatherland*, as having its own weather – the combined perspiration and breath of over 180,000 individuals condensing in the domed roof to form clouds.

At the southern end of the Avenue of Splendours was to be the second focal point of the city – a triumphal arch so large that the whole of the Arc de Triomphe in Paris could sit inside its inner opening. At the time, it was debatable, because of their grand scale and the restrictions imposed by the square-cube law, whether either of these constructions – the Volkshalle or the triumphal arch – would be realisable on Berlin's somewhat soft and unstable ground. To test the feasibility, in 1941, Speer constructed a concrete cylinder, 21 metres across and 14 metres high, weighing 12,650 tonnes. This was mounted atop an 18-metre tall and 11-metre diameter concrete foundation, which was embedded in the ground. The whole construction was then monitored for sinkage. If it were to have sunk less than 6cm into the ground, it would have been deemed sufficiently solid to allow the construction of the triumphal arch without further reinforcement. As it turned out, the cylinder sank nearly 20 centimetres in just two and a half years. The nonlinear square-cube law meant that the proposed triumphal arch, at three times the size of the Arc de Triomphe, would have 27 (3^3) times the mass, but only 9 (3^2) times the base surface area, consequently exerting three times as much pressure on the ground supporting it.

There is some poetic justice in the notion of this concrete cylinder, symbolising Hitler's desired but ultimately unrealised triumph, sinking down into a hole of its own making.

A nonlinear world

I have started playing card games with my kids. They've got to grips with the make-up of the pack – the ten number cards and the three face cards in each of the four suits. We play some simple games like beggar-my-neighbour, twenty-one, whist and rummy. With twenty-one in particular, they're at a stage where they are building strategies to play the game. They understand that eleven is a good total to be dealt with your first two cards. In twenty-one, there are four times as many cards in the pack with the value 10 (10s, jacks, queens and kings) as there are cards of any other rank. Roughly speaking (although this, of course, depends on the cards already dealt), starting from eleven, it's four times as likely that you'll hit twenty-one with your next card than it is starting from any of the numbers twelve to twenty. To put it more simply, you're four times as likely to draw one of the sixteen cards worth ten from the pack than to draw any of the four cards with any other single value. If the rules were different, so that only 10s, jacks and queens were worth 10, with kings worth 11, perhaps, then, you would only be three times as likely to draw a card worth ten and so on.

This is precisely the essence of linear direct proportionality. The probability scales directly with the number of different card ranks we are hoping to select. We encounter direct proportionality everywhere. If, instead of playing cards, I'm baking with the kids and we want to make twice as many cupcakes as the recipe suggests, then we need to use twice as much of each of the ingredients. The ingredients combine linearly to make twice as much mixture. This seems only right. It wouldn't make sense if we had to use three times the ingredients to make twice as many cakes. With direct proportionality the whole is no more or less than the sum of its parts. Double the parts and you double the whole.

But to propose this should hold true of every phenomenon in our world would be to deny the existence and the magic of emergent phenomena – the wetness of water possessed by no single molecule of H_2O, the beautiful murmurations of starlings impossible for one bird to choreograph on its own, the unique fractals of snowflakes formed, not by adding individual crystals together, but as one complex superstructure, and the intricate complexity that is the essence of all life on earth itself – lives which are so much more than the simple sum of atoms and molecules which comprise their physical embodiments.

Although most of the time we are unaware of them, many of the most important relationships that we experience every day are non-linear. But we have the idea of linearity drilled into us so early on and so often that sometimes we forget that other relationships can even exist. This indoctrination is so pervasive that, as we saw in iterated function learning experiments, our subconscious expectation of the relationship between two variables is that they are directly propor-tional to each other. Our overfamiliarity with linear relationships means that when something occurs that is nonlinear, it can catch us off guard and confound our expectations. By making the implicit assumption that inputs scale linearly with outputs, we are liable to find that our predictions can be way off the mark and that our plans can blow up in our faces. We live in a nonlinear world. Our brains, however, are so used to thinking in straight lines that we often don't notice it. We impose our linear view on every situation, assuming, as time goes on, that things will continue to change at roughly the same rate as they are now or that twice the effort will always garner twice the reward.

Of course, a more optimistic mindset with which to view our over-reliance on our straight-line expectations is that when they can be confounded in low-stakes situations, we can enjoy the surprise results. We often relish the astonishing revelations that occur when the linear scales fall from our eyes and our ideas are remoulded. To put a positive spin on it, our preconception that relationships between two variables can be depicted using straight lines creates a space for wonder and surprise which wouldn't exist if we could perfectly perceive how every nonlinear scenario would play out in advance.

When the kids and I are done with cards and it's time to clear up after baking, if I'm feeling mischievous, I will occasionally cheat by playing on the kids' linear expectations. I ask them to name any two ranks of card in the deck, from ace through to king. Let's say they choose queens and fives. I tell them that I will shuffle and if, when we go through the deck, we find any instances of those two ranks next to each other, in this case a queen next to a five, it will be their job to stack the dishwasher. If not, I will do it. Picking just two ranks – eight cards in total – I think their expectations are, quite reasonably, that the probability of finding cards of the two chosen ranks next to each other in the shuffled pack is pretty low. In fact, the probability is surprisingly high at roughly 50 per cent. Half the time, despite their initial expectations, they end up stacking the dishwasher. Just as with the birthday problem we met in Chapter 2, the maths is about pairs of cards, rather than individual cards. The nonlinear relationship means there are sixteen possible pairings between the four cards of each of the different ranks – the four fives and the four queens. With each pairing able to come up in two different orders, the probability of finding them next to each other is more likely than you might think at first. My kids are always amazed when we go through the deck and find such a pair – if not entirely pleased about the consequences.

In this chapter, we have come across some of the everyday nonlinear relationships, from pizza value for money and sun-protection factors, to gain and loss relationships on the stock market, which can confound our expectations and lead us to draw the wrong conclusions or make the wrong predictions. In the final three chapters, we will see that the verbal linear arguments we rely on to understand the world around us may cause us to neglect nonlinear possibilities, leading to backfirings, feedback loops, chaos and a litany of other nonlinear surprises.

7

DODGING SNOWBALLS

In the first months of 2020, the novel coronavirus, SARS-CoV-2, spread from China's Hubei province to countries around the world. The Italians isolated their first coronavirus cases on 29 January. By 11 March, Italy was reporting the highest number of infections outside China. Much of the rest of Europe and the West watched with horror the situation unfolding in Italy, hoping desperately that the same fate would not befall them. But one by one, almost every country began to detect their own cases. The pattern was usually the same. The numbers started off small – one or two people in the first reports, then a handful, often linked directly or, more worryingly, indirectly to the first, then it snowballed from there.

By 9 March, the situation in Italy had already become so serious that Prime Minister Giuseppe Conte was forced to impose a national lockdown. By mid-March, many Italian hospitals were overwhelmed, unable to cope with the huge volumes of Covid patients requiring treatment. Many in the UK watched on in sorrow and disbelief, but not with the absolute certainty that the same scenario would play out in their own backyard in a few short weeks. The numbers of cases reported in the UK were, after all, much, much lower than in Italy. Perhaps, the UK public hoped, we might escape relatively unscathed. Certainly, based on the Covid statistics in the UK, the government saw no reason in mid-March to institute a national lockdown to curb the spread of the virus.

On 15 March, the seven-day average for Covid deaths in Italy was up at 206. On the same day, the seven-day average for Covid deaths in the UK rose to just fifteen, up by five from the day before. Given these small numbers, it was still hard to imagine how the UK could soon find itself in the same situation that Italy was facing. Even if the UK's daily death toll continued to increase by the same amount (five) every day, it would take well over five weeks for its seven-day average death rate to match Italy's.

To many, there seemed to be no urgent need to implement an economy-damaging lockdown in the UK. Indeed, some argued that the numbers were so low that it might have been hard to persuade the general public of the need for such drastic measures. Perhaps we could wait to see how things would pan out? However, evidence from other countries, like Australia and New Zealand, suggests that the catastrophe does not have to be unfolding on one's own doorstep for a population to take lockdown seriously.

In reality, deaths were not growing linearly – increasing by the same amount each day – as the last chapter suggests we have been pre-conditioned to assume. They were growing *exponentially*, reflecting earlier exponential growth in cases. The UK surpassed Italy's daily death figures of 15 March just twelve days later, recording a seven-day average of 240 deaths on 27 March.

Exponential growth is yet another of the nonlinear phenomena that many of us fail to grasp intuitively. The mathematical definition says that a quantity that increases with a speed proportional to its current size will grow exponentially. This means that as the quantity increases, so does the speed at which it grows. The more infected people we have in the early stages of a disease outbreak, for example, the more people they will infect and the more the number of cases will rise. Other situations in which exponential growth plays a critical role range from pyramid schemes (the number of new investors increases in proportion to the number of investors already in the scheme) to nuclear weapons (the number of fissioning uranium atoms increases in proportion to the number currently splitting).

Underestimating the speed of growth of exponential processes – assuming they will grow more slowly than they do in reality – is known as *exponential-growth bias*. For many, it is a form of the linearity bias we met in the previous chapter; growth is assumed to be linear when it is, in fact, exponential. Studies into the phenomenon have shown that higher levels of income or education make little difference to people's ability to avoid falling into the trap of underestimating exponential growth.[103] Indeed, it is common, even for people who have previously encountered exponential growth in situations like the calculation of compound interest, to fail to recognise the phenomenon in other contexts.[104]

In one study from 2016, economists Matthew Levy and Joshua Tasoff presented subjects with questions like 'Asset A has an initial value of $100 and grows at an interest rate of 10 per cent each period and Asset B has an initial value of $X and does not grow. What is the value of X which would make the two assets equal after twenty periods?' and asked them to rate how confident they were in their answers.[105]

Being able to answer such questions accurately (or otherwise) has a significant impact on the soundness of our financial decisions today with a view to their ramifications in the future. Get your calculator out and have a go for yourself. See how confident you feel at the end of your calculation.

To find the correct answer we need to increase the total of Asset A by 10 per cent in each of the twenty periods. This involves multiplying the initial $100 by 110/100 or, simplifying, multiplying $100 by 1.1, twenty times (equivalently calculating $100 \times (1.1)^{20}$). Plugging this into the calculator gives the value of X to be 672.75. Even with a calculator to hand, the majority of subjects in the experiment did not get the answer correct. A third gave the answer $300, which is exactly what would have happened if the growth had been purely linear and growing by the fixed amount of $10 (10 per cent of the original investment) at each period – those subjects were thinking too linearly. Perhaps the most startling aspect of the study was that the people who performed worst on the test were also those who were most confident in their answers.[106] Many of us don't even know that we don't know about exponential-growth bias.

The ramifications of underestimating the long-term impact of compound interest can have severe financial consequences. Consumers underestimate how quickly a given amount of money compounds and consequently underestimate its future value. This makes saving seem less attractive. Individuals undervalue the utility of investing for their future, leaving themselves without proper provision for their old age.[107] Miscalculating exponential growth also makes taking on debt more attractive, as people underestimate the scale of the repayments. Real-world studies of the phenomenon suggest that the effect of exponential-growth bias might be to double an individual's debt-to-income ratio[108] (the amount of debt they take on relative to their income) in comparison to someone who does not suffer from the bias.

This inability to correctly recognise and interpret exponential growth has also been shown to act as a significant impediment to the implementation of effective strategies to control infectious disease.[109] One study from 2020, focusing on the early stages of the pandemic, found that the higher the exponential-growth bias exhibited by a subject, the lower the levels of compliance with anti-Covid measures like the use of face coverings and social distancing. People who were unable to accurately estimate the speed of disease spread were unable to see the importance of disease-control mitigations, and hence less likely to implement or observe them.[110]

President Trump may have been a high-profile example of the failure to understand exponential growth. He placed huge emphasis on the low absolute number of cases in the United States in the early stages of the pandemic, not appearing to recognise how quickly these numbers could take off. Consequently, the Trump administration continually downplayed the seriousness of the situation, which, in turn, led to reluctance to implement the mitigations that were required to bring the virus under control.

'So last year 37,000 Americans died from the common flu. It averages between 27,000 and 70,000 per year. Nothing is shut down, life & the economy go on. At this moment there are 546 confirmed cases of CoronaVirus [sic], with 22 deaths. Think about that!' Trump tweeted on 9 March 2020. The figures Trump quoted for flu were exaggerations. According to the Centers for Disease Control and Prevention (CDC), during the flu season of 2018–19, roughly 34,000 people died from influenza. But the CDC also put the average number of annual flu deaths since 2010 at somewhere between 12,000 and 16,000 – much lower than Trump had claimed. Although his Covid figures were roughly correct (on the day he made the announcement the US had seen a total of 594 confirmed cases and 22 deaths) he failed to appreciate just how rapidly the situation would evolve.

By the time his presidency ended on 20 January 2021, despite some of the attempted mitigations that had been put in place, there had been a total of 24.5 million Covid cases recorded in the US and over 400,000 deaths, dwarfing even his own exaggerated flu death figures. Indeed, the same study that highlighted the importance of understanding exponential growth for the observance of anti-Covid mitigations also found that conservatives in the US were more prone to underestimating the absolute growth rate of the epidemic than liberals.[111]

The positive news unearthed by this study, however, was that compliance with anti-Covid mitigations could be improved dramatically through different data representations, allowing people to see the scale of the growth using raw numbers, rather than graphically. When people better understood the true rate of growth of the epidemic, their perception of risk was heightened and they were more likely to comply with suggested protective behaviours.[112]

Even people who recognise when a process is subject to exponential growth are prone to underestimating the potential for rapid change. Indeed, this is exactly what happened in the UK in March 2020, as we watched the situation in Italy unfold. Although the scientists informing the government recognised that the growth of cases would be exponential, they hugely underestimated the speed of the growth.

On 12 March, in a live broadcast from Downing Street, the people of the UK were told that, 'On the curve, we are maybe four weeks or so behind [Italy] in terms of the scale of the outbreak'. At that point, the UK had a total of 590 reported cases, whereas Italy had over 15,000. There were just two new Covid deaths reported on 12 March in the UK, while in Italy there were 189. With such a huge disparity in the case numbers and a low number of daily deaths, many people unquestioningly believed that we had a four-week head start.

There is a common misconception that exponential growth means fast growth. But it doesn't always. At the early stages of an epidemic, exponential growth can be disarmingly slow. When case numbers are low, so is their growth. But things can get out of hand incredibly quickly. This is especially dangerous if, for example, you think you are further behind the curve than you are.

Perhaps the most crucial figure in understanding how fast an exponential process is growing is the so-called 'doubling time'. In the early stages of an epidemic, this is the time for cases, hospitalisations or deaths to increase by a factor of two. The consistent doubling of these statistics in a fixed period of time is the hallmark of exponential growth. On 16 March, Boris Johnson told the press that '. . . without drastic action, cases could double every five or six days'. This figure is reflected in the minutes of the UK's Scientific Advisory Group for Emergencies (SAGE), the scientific body that was advising the government on the pandemic. In their minutes from 18 March, they quote a doubling time of '5–7 days'.

This doubling time explains where the 'four weeks' figure comes from. With a doubling rate of six days (in the middle of the SAGE estimate), the time to get from the UK's total of 590 reported cases to Italy's 15,000 (both reported on 12 March) would have been twenty-eight days – exactly four weeks.

But this five-to-seven day doubling time was wrong. Drastically wrong.

A more accurate doubling time in the early stages of the UK's epidemic has been calculated to be around three days. Although SAGE's estimate of the doubling time is only out by a factor of two, which doesn't sound too bad, the undeniably dramatic exponential spread of the disease means that this error is compounded – itself doubling every few days. The three-day doubling time predicts that the UK would have reached Italy's 12 March total of 15,000 cases around two weeks later. This estimate was borne out in reality, with the UK hitting 17,000 cases on 28 March – just sixteen days down the line.

The potential ramifications of the UK government thinking they had more time than they did, and that the epidemic was growing more slowly than it was, were huge. This false sense of security might have contributed to the delay in taking measures to suppress Covid, which resulted in the avoidable loss of tens of thousands of lives during the first wave of the UK's epidemic.

Positive feedback loop

The exponential growth seen at the beginning of a disease outbreak is an extreme example of a more general phenomenon – the *positive feedback loop*. Positive feedback loops are characterised by a signal, which triggers a response, or a series of responses, which ultimately end(s) up amplifying the original signal, closing the loop. In an epidemic, for example, infected individuals can come into contact and infect susceptible people, creating more infectious individuals who have the power to go on and infect more people and so on.

Positive feedback loops can amplify an initially small quantity to unexpected magnitudes. For this reason, the impact of positive feedback is sometimes referred to as the *snowball effect*. In the analogy, a small amount of snow that begins rolling down a hillside picks up more snow as it rolls and increases in size. The bigger it gets, the more snow it picks up, until the initially small snowball has become huge and uncontrollable. The snowball itself, however, is only ever a metaphor. When positive feedback acts on real snow falling down a mountain, the result is not an enormous snowball but, often, a deadly avalanche carrying tens of thousands of cubic metres of snow down the hillside. Sadly, residents of the mountainous Kinnaur district of Himachal Pradesh in northern India are only too familiar with the effects of positive feedback.

*

July marks the beginning of monsoon season in Himachal Pradesh. On 17 July 2021 the Indian Meteorological Department issued orange alerts for rain across the region. Despite well-publicised warnings to tourists about the dangers of visiting Himachal Pradesh in monsoon season, many had flocked to the region in July after the relaxation of some of India's Covid restrictions. One of these eager tourists was Dr Deepa Sharma, a well-respected women's rights advocate, visiting the region from Jaipur. At 12.59 on 25 July, she shared with her tens of thousands of Twitter followers a selfie taken at Chitkul, the last occupied Indian village before the border with Tibet. Less than half an hour later, the results of a positive feedback loop had cost her her life.

A small, seemingly innocuous ground vibration, a gust of wind or a drop of water on the surface of a mountainside may loosen the topsoil, which then shifts a pebble, which, in turn, hits a stone and causes it to displace a bigger stone, which leads to increasingly large rocks surging down the mountainside. The vibrations from these falling boulders may further loosen the surrounding ground, causing more debris to cascade, until the valley below is showered by a deadly rockslide. The initial signal – a tiny amount of mass displaced on the hillside – can be amplified dramatically by the positive feedback loop. Heavy rain substantially increases the risk of rock and landslides by lubricating and adding weight to the uppermost layers of soil.

Despite it not having rained on the day of Deepa's trip to Himachal Pradesh, heavy rain in preceding days meant that the hillside above the Kinnaur's Sangla Valley was primed for a rockslide. Mobile-phone footage of the slide shows enormous boulders crashing down the hillsides, some occasionally catapulted into mid-air by a small up-ramp on the hillside, seemingly suspended for impossible amounts of time before crashing into the Baspa river below. One boulder on its descent can be seen to smash directly through the stainless-steel road bridge spanning the river, as if it were made of matchsticks. At 13.25, a boulder, released as an end product of the positive feedback loop which caused the rockfall, smashed into the sightseeing bus in which Deepa was travelling, killing her and eight of her fellow passengers instantly.

*

Positive feedback loops can have real and deadly consequences when played out at a large enough scale. One of the deadliest catastrophes our species will likely ever experience, anthropogenic global warming, is causing extreme weather conditions, falls in food production and changes in disease-transmission patterns that are already purportedly costing 150,000 lives a year. One of the most worrying aspects of global warming is that much of the predicted temperature rises are already locked in through the impact of positive feedback loops. One such loop is known as *ice-albedo feedback*. Albedo refers to the amount of solar radiation hitting the earth that is reflected back into space. Glacial, sheet and sea ice, being white, tend to reflect a high proportion of the radiation that falls on them. When the global temperature rises, some of this ice starts to melt. This melting alters the earth's albedo, exposing a higher proportion of the considerably darker land and sea, which absorbs more of the sun's radiation. This raises the temperature, which then causes more ice to melt, which further lowers the albedo and so on. Positive feedback means that even acting to reduce carbon emissions dramatically today, it is likely that we will still see global temperatures rise by at least 1.5 degrees.[113]

More innocuously, many of us have experienced the high-pitched screech of a microphone when it is in close proximity to a speaker. This 'audio' or 'acoustic' feedback loop is the result of the microphone picking up a signal which then gets relayed to the speaker. The microphone picks up the amplified sound sending it back to the speaker. And so the loop continues. Although most of us associate audio feedback with a high-pitched whine, it's possible to achieve feedback at much lower frequencies. The frequency that ends up getting amplified most strongly, and hence the one that dominates what we hear, depends on the relative positioning of the speakers and the microphone, as well as the natural acoustics of the room and the properties of the speaker itself. Many speakers produce high-frequency sounds more efficiently than lower frequencies, which is why we tend to hear these wince-inducing tones, rather than the deeper ones.

In the 1960s and 70s the deliberate use of electric-guitar feedback enhanced the range of tools that artists like the Grateful Dead, the

Velvet Underground, Jeff Beck and the Who could employ to create their distinctive sounds. The most prominent exponent of this positive feedback effect was probably Jimi Hendrix, the distorted sounds he was able to produce being almost synonymous with much of his work.

Perhaps more abstractly for most of us, we also observe these sorts of positive feedback loops in the form of stock-market bubbles, like those we encountered in the previous chapter. In this case, the feedback loop has two major components: the investors, and the prices of the stocks and shares in which they invest. Investors who are influenced by price movements take actions which themselves influence the price (note that the companies and their performance play a worryingly small role in such price movements). Imagine a company that has been performing strongly and has therefore been able to offer its shareholders a dividend. This may, quite naturally, cause the share price to increase. This movement in the share price may draw in more investors for the limited shares on offer. Demand may outstrip supply, which causes the share price to rise higher. This, in turn, draws in more investors chasing the spiralling price. The problem, as we saw in the previous chapter, is that feedback loops can decouple a company's share price from its performance. Small fluctuations in price can be amplified far beyond the scale that their cause merits. This has the potential to leave companies, and indeed whole sectors, massively overvalued.

All the examples of positive feedback we have met in this chapter thus far (disease spread, rockslides, global warming, acoustic feedback and financial bubbles) have involved the often rapid or out-of-control increase of one quantity or another (namely infected individuals, descending rock mass, temperature, volume or share prices). It's tempting, therefore, to associate the *positive* in 'positive feedback' with the *increase* of the quantity concerned. However, we should correct this misconception and warn that positive feedback, as illustrated in Figure 7-1, can lead to your share prices going down as well as up!

Although the ice-albedo feedback loop may be responsible for some of the recently observed rises in global temperature, in the more distant past, it has also been responsible for huge falls in temperature. The *Snowball Earth hypothesis*[114] suggests that around 650 million years ago, the whole (or almost the whole) of the earth (including the oceans) was covered in a layer of ice. Adherents to the theory suggest that falls in the earth's temperature led to increases in the amount of sea ice, which increased the earth's albedo, sending more of the sun's radiation back into space and further cooling the temperature of the planet. This cooling spiral is another example of a positive feedback loop, but one that operates in the opposite direction to the one which plays such a significant role in global warming today. The 'Snowball Earth' name given to the hypothesis is doubly appropriate, then, describing both the end point of the process – a giant ball of ice and snow – as well as the snowball-effect positive feedback loop which caused the deep freeze.

The positive feedback loop which causes share prices to rise in the stock market can also run in the opposite direction. For example, when enough individuals think a share is overpriced, people will be trying to sell more shares than others are willing to buy – supply will outstrip demand. This excess offering of shares will cause the share price to decrease, which, in turn, may cause more people to cut their losses by selling, leading to a larger surplus of offered shares and lower prices.

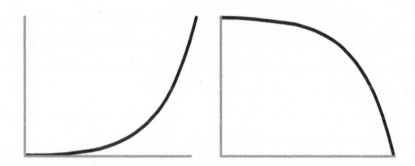

Figure 7-1: Quantities subject to positive feedback can both grow (left) and decline (right).

*

Positive-feedback loops can also have powerful consequences in other financial contexts, which reach beyond the world of high-flying investors into the lives of ordinary people. In the summer of 2007, the UK bank Northern Rock was a source of economic pride for many of the people of the north-east of England. After decades of declining industry and high unemployment, 'The Rock' offered the real possibility of carving out a lucrative share of the financial-service market for the north-east, and the ability to compete in this London-centric sector. Through investing in local communities and charity initiatives, Northern Rock's identity was fast becoming synonymous with the north-east and the city of Newcastle in particular. The bank was the major shirt sponsor of a number of high-profile local sports teams, including Durham County Cricket Club, Newcastle Falcons Rugby Club and, most notably, Newcastle United Football Club.

The bank had levered itself into a prominent position in the FTSE index of one hundred leading UK companies through an aggressive programme of borrowing on the international money markets during the early 2000s. These loans allowed the bank to ambitiously expand its mortgage portfolio, becoming the UK's fourth-biggest bank by share of lending. The Rock's business model worked by reselling their UK-based mortgages on the international markets to generate money to pay off its loans. Although financially viable and not without precedent, the strategy relied on placing a large number of the bank's eggs in a single, potentially fragile basket.

As the subprime mortgage crisis in the US grew over the summer of 2007, international demand for Northern Rock's mortgages slumped. On 13 September, the bank was forced to apply to the UK government for short-term liquidity support – embarrassing and undesirable, certainly, but not in itself an unmitigated disaster for the business. Emergency funding was quickly granted, which meant, in theory, that Northern Rock was still a viable ongoing business. However, the publicity surrounding this unprecedented bailout from the Bank of England was what really did for the bank.

In the media reporting of this complex financial scenario, much of the subtlety was unsurprisingly lost. Many media outlets reported that the UK government was the 'Lender of Last Resort' – hardly a confidence-inspiring phrase. Experts pointed out that in the event of a collapse, current legislation would only protect savings of up to £33,000. The *Daily Mail's* headline on the crisis the next day was: 'Northern Rock: Greed and stupidity crocked the rock'. The clear message that reached members of the public with funds deposited in Northern Rock was that the bank was in serious, maybe terminal trouble.

The next day, queues of customers looking to withdraw their savings began to form in the streets outside branches of the bank up and down the country. The bank's website collapsed under the strain of increased traffic and their phone lines quickly became saturated. For a bank with liquidity problems, savers withdrawing their funds was the last thing it needed. The firm's shares lost 32 per cent of their opening value that day. Media reports of huge numbers of people attempting to withdraw their money, accompanied by anxiety-inducing images of the ever-growing queues, further fuelled the crisis. It quickly became understood that the bank would not be able to pay out all its depositors. Understandably, this did nothing to dampen the eagerness of savers looking to withdraw their cash, each one hoping desperately not to still be in the queue when the money ran out. Driven by more negative news headlines, the run on the bank continued for four more days. Despite the Chancellor of the Exchequer publicly announcing that the government would guarantee all Northern Rock's deposits, the bank never recovered. Thousands of heavily invested shareholders lost their life savings when the bank was nationalised in February 2008. The widely broadcast message that the bank was in trouble had become a reality.

A run on the banks is a special example of a feedback loop known as a self-fulfilling prophecy – a prediction, suggestion, belief or report, the reaction to which brings it to fruition. Mervyn King, then Governor of the Bank of England, said he would have liked to have offered Northern Rock financial support in secret to avoid the confidence-denting public announcement that led to its downward spiral. However, financial regulations didn't allow for such secrecy. Had the fact that Northern Rock was in trouble not become public knowledge, it is likely that the crippling bank run may never have occurred.

The name game

An *aptronym* is the term given to a name which is particularly suited to its owner, usually because it is related to their occupation or some other defining characteristic. An old friend and I have an email correspondence which is almost exclusively composed of examples of such people that we've spotted in the wild. Some of the more obvious ones that many will be familiar with are Usain Bolt, the world's fastest runner, Margaret Court, the former world-number-one tennis player and Thomas Crapper, the plumber and toilet designer who, contrary to popular belief, did not actually give an abbreviated form of his name to a slang word for defecation.

Possibly less well known, but arguably even more apt, are the Jamaican cocaine trafficker Christopher Coke, the British judge Igor Judge and the American columnist Marilyn vos Savant (who between 1985 and 1989 was listed in the *Guinness Book of Records* as having the world's highest IQ). Sara Blizzard, Dallas Raines and Amy Freeze are all television weather presenters, Russel Brain is a British neurologist and Michael Ball is a former professional footballer. I could go on. Some of these examples seem almost too apposite to have happened by chance. As we investigated in Chapter 2, when we come across unexpected connections like these, we are apt to assume a causative link – surely the reason why these people ended up being renowned for their particular speciality was due to the influence, from an early age, of the name they bore? The hypothesis that such causative links exist is another example of a self-fulfilling prophecy and is known as *nominative determinism*.

Although it's easy to dismiss the idea of someone's name having an impact on their vocation in life, many researchers would like to believe it. One proposed explanation for why people might be drawn to professions which fit their name is a psychological phenomenon known as *implicit egotism* – the conjecture that people exhibit an often unconscious preference for things associated with themselves. That might be marrying someone with the same birthday, donating to good causes with a name that begins with their initial or gravitating towards a job which relates to their name. In support of this idea, James Counsell mused on his eventual career path as a barrister, 'How much is down to the subconscious is difficult to say, but the fact that your name is similar may be a reason for showing more interest in a profession than you might otherwise'.

There are a limited number of studies which purport to provide evidence that nominative determinism is a real phenomenon.[115] Perhaps the most amusing of these[116] was conducted in 2015 by a family of doctors and soon-to-be doctors: Christopher, Richard, Catherine and David Limb. Together the four Limbs clearly had a vested interest in understanding whether their appendage-related name had drawn them towards their anatomically focused professions. Indeed, given the vocation of David Limb as an orthopaedic surgeon (specialising in shoulder and elbow surgery), the Limbs decided to ask a more particular question: whether a doctor's name could influence their medical specialisation.

By analysing the general medical council's register, they found that the frequency of names relevant to medicine and its subspecialities was far greater than would be expected by chance alone. One in every twenty-one neurologists had a name directly relevant to medicine (like Ward or Kurer), although far fewer had names relevant to that particular speciality (no Heads or Parkinsons, for example). The specialities next-most likely to have medically relevant names were genitourinary medicine and urology. The doctors in these subfields also had the highest proportion of names[117] directly relevant to their speciality, including Ball, Koch, Dick, Cox, a single Balluch and even a Waterfall.

As the Limbs pointed out in their paper, this may have had something to do with the wide array of terms that exists for the parts of the anatomy relevant to these subfields. Ironically, despite the purported evidence for the phenomenon, the fact that the two younger (lower?) Limbs followed their parents (upper Limbs?) into their profession hints at a strong role for familial influence in determining careers, in medicine at least.

In 2002, researchers at Montgomery College in Maryland scoured a variety of databases which linked names and occupations. One of their most interesting findings was an unusually high number of men named Dennis in the field of dentistry in comparison to other names in the same profession.[118] In order to demonstrate that this wasn't just because Dennis was a more popular name than many others, the authors also analysed the number of dentists called Walter – the next-most popular name in the 1990 census (and also coincidentally Dennis' arch-rival in the British cartoon strip *Dennis the Menace*). They found that Walter was a far less common name for dentists than Dennis and suggested nominative determinism, mediated by implicit egotism, as the underlying causative reason.[119] Critics argued that the comparison wasn't a fair one.[120] While both the names Dennis and Walter had decreased in popularity over the years, Walter peaked as a baby name in 1892, whereas Denis didn't peak until 1946. Despite being evenly represented overall on the 1990 census data, the Walters were significantly older than the Dennises, meaning you would be more likely to find men named Dennis[121] in any profession in comparison to men named Walter, who might now have retired.

Taking account of this criticism, the Maryland researchers went back to the drawing board.[122] Using census data from the US in 1940, they found that men in eleven professions – baker, barber, butcher, butler, carpenter, farmer, foreman, mason, miner, painter and porter – were, on average, 15 per cent more likely to work in occupations corresponding to their surname than random chance would dictate. The detailed census data also allowed them to rule out confounding factors like ethnicity and educational attainment as possible explanations. They found similar evidence for nominative determinism in the 1880 US census and the 1911 UK census.[123]

Before we decide whether we believe that our names can influence our future trajectories though, it's important that we remember some of the lessons we learned in Chapters 2 and 3. While the correlation between the eleven names was noteworthy, there were presumably plenty of other names like Archer, Taylor, Bishop and Smith, for example, that did not have a clear correlation with corresponding employment, otherwise they would have been included. The names that were highlighted could represent a form of reporting bias. It's possible that the over-representation of the eleven names listed in the study appeared purely at random and that drawing the target around them by highlighting them in the paper constitutes an example of the sharp-shooter fallacy we first encountered in Chapter 2. It is also important to remember that correlation does not imply causation. As we saw with the Dennis the Dentist example, there may be confounding factors, which were not controlled for, and which were influential in generating the correlation between the names and the occupations. Whether or not nominative determinism is truly a self-fulfilling prophecy, finding examples of aptronyms like the lawyer Soo You, the Washington news-bureau chief, William Headline, the pro-tennis player Tennys Sandgren or the novelist Francine Prose still makes me smile.

Self-fulfilling prophecies are also popular literary tropes. In antiquity, they were commonly used to illustrate *inexorable fate* or destiny – the concept of a predetermined, inescapable future. This predestination is fundamental to the Hellenistic world view, which is perhaps why self-fulfilling prophecies turn up so often in ancient Greek myths and legends. Probably the most famous example from this civilisation is the story of Oedipus Rex.

Before he is born, the Oracle at Delphi makes a prediction that Oedipus will kill his father and marry his mother. Hoping to avoid this fate, his parents, the King and Queen of Thebes, leave their newborn to die on a hillside from where he is eventually rescued and raised by another family. As a grown man, feeling unsettled, Oedipus decides to consult the same Oracle, who visits upon him the same prophecy. Believing his much-loved foster parents to be his real parents and hoping to avoid the predicted fate, Oedipus decides to travel to Thebes to start a new life.

On the way, he fights with and kills a stranger he meets on the road who, it later transpires, was his real father, thus unknowingly fulfilling the first part of the prophecy. When he arrives at Thebes, he rids the citizens of the Sphinx, a terrifying monster that has been plaguing the city. As a result, he is rewarded with the hand in marriage of the recently widowed Queen of Thebes, his own true mother. After they marry, completing the second part of the prophecy, the terrible truth eventually comes out and the news of their unintentional incest doesn't go down well with either Oedipus or his mother. Most of the protagonists of the story end up miserable (or dead), apart from the smug Oracle, who by their very intervention, manages to make yet another of their predictions come true. So influential is this story, that famous philosopher of science Karl Popper referred to self-fulfilling prophecies as the 'Oedipus effect' in his early and influential work.[124]

More recently, the self-fulfilling-prophecy device was employed by J. K. Rowling in her best-selling *Harry Potter* series. Voldemort, the arch-villain of the franchise, becomes obsessed by a prophecy made at the time of Harry's birth: that a child matching Harry's description would have the power to bring about his downfall. Not long after his birth, Voldemort attempts to kill the infant Harry, but fails, killing his parents instead and, in the process, instilling in Harry a strong desire for revenge. During his failed assassination attempt, Voldemort also accidentally transfers some of his power to Harry, which (without giving too much away) makes the boy an exceptional wizard, giving him the means to eventually dispense with Voldemort.

Predestination paradoxes, primarily exploited in science fiction, are like self-fulfilling prophecies shifted backwards in time. In an attempt to change something in the recent past, the protagonist travels further back in time, but eventually ends up causing the event they were trying to avert. For example, you might go back in time to prevent the burning of the Great Library of Alexandria. While sneaking around in the dark to avoid being seen, you might accidentally knock over a lamp which ignites the fire which motivated the time travel in the first place. These sorts of literary devices are designed to illustrate the fundamentally unchangeable nature of *past* events.

In the film *12 Monkeys*, protagonist James Cole is sent back in time to prevent the release of a deadly virus that wipes out huge swathes of the human race. Unfortunately, instead of wiping the virus out, Cole's chance meeting with a burgeoning anti-establishment scientist seeds the anarchist idea of unleashing the virulent holocaust on the world. Cole's travels in time end up causing the fulfilment of the very events he is trying to prevent.

Psychologists distinguish two types of self-fulfilling prophecies which can influence individuals: *self-imposed* and *other-imposed*. In self-imposed prophecies, one's own expectations are the initiator of the feedback loop which drives the series of events that make those expectations a reality. For example, a study from the University of British Columbia found that when people perceived they were the victims of gossip or the recipients of snubs in the workplace, they were more likely to undertake actions (like eavesdropping or spying), which increase the likelihood of rejection by colleagues.[125] The employee's own paranoia was the catalyst for bringing about the expected spurning.

As well as our social interactions, self-imposed self-fulfilling prophecies can also have an impact on our health. When chatting recently with a Chinese friend over lunch, the subject of birthdays came up. I remarked that the repeated digits in my birthday – 4 April – made it particularly easy to remember. Upon hearing this date, her face dropped, and she offered me commiserations for being born on such an inauspicious day. I admitted to her that I had no idea that my birthday was unlucky and that I had always considered it rather a special day, for obvious reasons. She proceeded to explain that in China, and indeed many other East Asian countries, tetraphobia – fear of the digit 4 – is a relatively common superstition. In part, this is because the word for four sounds similar, if not identical to the word for death in many of these countries. This superstition pervades many cultures, to the extent that floor numbers often skip from three to five in tower blocks or hospital buildings, while military aircraft and ships eschew the number in their names and table numbers at weddings bypass the digit 4. My birthday – on the fourth day of the fourth month – is considered to be extremely unlucky and is often avoided for meetings and appointments.

So ingrained is the phobia that when US mortality statistics between 1973 and 1998 were studied, researchers found that Americans of Chinese and Japanese heritage were more likely to die of heart problems on the fourth day of each month than on any other day.[126] Controlling for a range of potentially confounding factors demonstrated that the increased stress, experienced among these populations as a result of the widespread superstition about the number 4, led to higher rates of death, fulfilling the 'unlucky-number-4' prophecy.

The lie that makes itself come true

The most famous self-fulfilling prophecy in the medical arena is the *placebo effect*. A placebo is a drug or procedure that appears to be a real medical treatment but isn't. Sugar pills are a classic placebo; they can be made to look like real medicine, but typically will not contain substances which have any therapeutic benefit. The effect occurs when the placebo treatment actually improves a patient's symptoms or makes them feel better, in part as a result of a self-fulfilling expectation that the treatment will work. As a result, the placebo effect is sometimes referred to as 'the lie that makes itself comes true'.

In the late eighteenth century, American physician Elisha Perkins developed a medical device he called the Perkins tractors. He claimed that the pair of metal rods could heal people of various ailments, including inflammation, rheumatism and head pain, by drawing pathogenic 'electrical fluid' away from the body. Indeed, the people on whom Perkins used his rods reported feeling benefits from his prodding and wafting. On the basis of the improvements that his device seemed to produce, Perkins was issued a medical patent for the Perkins tractors in 1796. The President of the United States, George Washington himself, is said to have been sufficiently impressed to have purchased a set (although given his death was hastened by bloodletting – another pseudoscientific therapy – it's dubious, in hindsight, how much his patronage is really worth). Whether the benefit that Perkins' patients were feeling was a genuine effect of the rods' healing powers, or the effect of the patients believing their conditions would be improved and their expectations being self-fulfilled, was not entirely clear. There was no way of quantifying the self-fulfilling placebo effect.

Because it is now well known that just the impact of being treated can alone produce a measurable benefit, clinical trials of new therapies will typically have a treatment group and a control group. Patients in the treatment group are administered the therapy under investigation, while those in the control group are administered a sham or placebo version of the same therapy. The placebo version looks and feels the same, but lacks the key property or active ingredient that the true therapy is based on. The difference in outcome between the treatment group and the control group can be ascribed to the effects of the therapy itself, rather than the self-fulfilling expectation induced by the ceremony surrounding the remedy, which both groups experience. If the patients in the treatment group show no difference in response when compared to those in the control group, then the therapy is no better than a placebo and has no discernible benefit.

If no placebo were administered in the control group, the patients in the treatment group may have shown a benefit in comparison to those patients receiving no treatment at all, but it would be impossible to tell whether this was simply due to the treatment group's reaction to the theatre of being treated or because of a genuine benefit. This is exactly what was happening with Elisha Perkins' tractors.

When Perkin's new devices crossed the Atlantic, they quickly became extremely popular, selling for what, at the time, was an extortionate sum of 5 guineas (over £500 in today's money). Bath – the city which later became home to my university – was already well renowned for its 'alternative' treatments, including its 'healing' mineral waters and spas, so the tractors found a natural market and became extremely popular in the city. However, Dr John Haygarth, one of the city's residents, was extremely sceptical of these new disease-conducting rods. To determine the true impact of the Perkins tractors, Haygarth developed a set of sham tractors made of wood but painted to look exactly like the real thing. The wooden tractors would provide the look of the original treatments but, not being able to conduct electricity, couldn't possibly act through the same mechanism.

In his trial, Haygarth treated five individuals with the 'real' metal tractors (the treatment group) and five with his own dummy version (the control group). He found that the fake wooden tractors were just as effective as the real ones. Interestingly, he didn't conclude that the tractors didn't make a difference. In fact, he found that patients treated with both the real and the fake tractors seemed to report a remarkable improvement (although notably he did not include in his trial a control group who were given no treatment at all for comparison). He later extolled in his book 'to a degree which has never been suspected, what powerful influence upon diseases is produced by mere imagination' – the first characterisation of the self-fulfilling placebo effect.

Placebo effects can be self-imposed or other-imposed or a combination of the two. For the self-imposed placebo effect the subject's own expectations are the catalyst for their improvement, despite the fact that the treatment they are taking is known (although not by them) to have no benefit. The self-imposed aspect of the impact diminishes if the patient is aware that they are being given the placebo treatment. For that reason, the patient must be 'blind' to which group of the trial they are in. This is why it was so important that Haygarth's dummy tractors looked exactly like Perkins' real ones. Any difference would have given the game away and alerted the patients undergoing the sham treatment, potentially shattering the illusion and diminishing the strength of the placebo effect in that group.

However, given the different densities of wood and metal, what Haygarth couldn't eliminate was the fact that he himself knew which tractors were real and which were his own sham version. And it turns out that this matters because of the 'other-imposed' aspect of the placebo effect. In a so-called single-blind trial, the subject doesn't know which group they are in, so their expectations cannot influence the result. But if the researchers carrying out the trial know, they may unconsciously (or worse, consciously) influence their patients, biasing the results in one direction or another. The subconscious influence of an experiment's participants by a researcher is characterised by the *observer-expectancy effect*. Perhaps the most remarkable example of the effect is the miraculous-seeming story of Clever Hans.

Too clever by hoof

At the turn of the eighteenth century, there was a great interest, in both the public and scientific spheres, in the extent and limits of animal intelligence. The foremost animal intellect of the day was alleged to be a German horse by the name of Hans, owned and trained by a former mathematics teacher and mystic Wilhelm von Osten. Von Osten claimed that Clever Hans had advanced abilities with numbers, being able to add, subtract, multiply, divide, manipulate fractions and tell the time, as well as being able to read, spell and understand German. Von Osten backed up his claims with public exhibitions in which he put Clever Hans to the test. He might ask him questions like, 'What is six multiplied by two?' In response the horse would tap his hoof twelve times. Having seen many variations of these sorts of questions answered correctly, almost all the invariably large audiences would go home thoroughly convinced of the animal's extraordinary brainpower.

The German Board of Education set up a commission to investigate von Osten's claims scientifically and to determine whether Hans really was 'clever'. Remarkably, the commission found that Clever Hans *could* give the right answer, even if his trainer was substituted for someone else asking the questions. This demonstrated that there was no underhand trickery being employed deliberately by von Osten to fool his audiences and seemed to point in the direction of Hans genuinely being a horse with human-like intelligence. However, in follow-up experiments, the commission found that Hans was only able to give the correct answer when the human asking the question knew the correct answer as well. Hans got only 6 per cent of the answers correct for questions to which von Osten did not himself know the answers.

After evaluating Hans' ability, the commission turned their attention to von Osten. They found that entirely unbeknown to the trainer himself, as Hans approached the correct number of hoof taps, von Osten would change his facial expression and posture, becoming almost imperceptibly more tense at each tap, until the correct answer was reached, and the tension abated. Hans could sense when von Osten expected him to stop. When the horse was fitted with blinders, so that he could no longer see his trainer, he was unable to answer the questions. Entirely unknowingly, von Osten had been feeding Hans the answers all along. Perhaps unsurprisingly, because he never realised he was doing it in the first place, von Osten continued to believe in Hans' abilities and exhibited him around the country, drawing in even bigger crowds.

There is evidence to suggest that the same phenomenon may be at play with sniffer dogs.[127] When searching members of the public for drugs, for example, their human handlers may give subtle unconscious signals indicating their underlying biases about a suspect and potentially leading to false alarms.

In the early 1960s, psychologists Robert Rosenthal and Kermit Fode were interested in whether laboratory animals could be bred to be smarter.[128] For the rats that were the subject of their study, intelligence was assessed by how quickly a rat could navigate a maze to find a food source. By the time they were ready to perform their decisive experiment, they had rats in one cage labelled as 'maze-bright' – the descendants of a long line of the smartest, most intelligent rats, who had been fastest at solving the maze in each generation. Rats in a second cage were labelled as 'maze-dull' – the progeny of the slowest, most intellectually sluggish rodents in preceding generations. When a group of student assistants put the rats through their paces, they did indeed find that the 'dull' rats recorded the slower maze-solving times and that the 'brighter' ones posted the faster times. More interestingly, the students handling the clever rats reported them to be more likable, easier to handle, and more fun to pet and play with. The findings could have raised some interesting questions about the links between intelligence and temperament, but it was never possible to follow them up.

The twist in the tale, as you may have already guessed by now, was that there were, in fact, no differences between the rats in the two cages at all. The whole supposed breeding programme was a sham. The rats were selected at random by Rosenthal and Fode, and placed in cages labelled to give the student assistants the impression that there was a real difference in ability. The difference in performance between the two groups was purely an artefact of the predisposition of the handlers. This is an example of what Rosenthal and Fode called the *experimenter effect*, an other-imposed self-fulfilling prophecy.[129] When you expect intelligence, intelligence is what you see, even in laboratory animals. Imagine how powerful the effect could be if it were harnessed for humans.

A few years after his landmark rat study, Rosenthal teamed up with elementary-school principal Lenore Jacobson to carry out another pioneering experiment, this time on Jacobson's schoolchildren and their teachers.[130] They first devised a fake test which they suggested to teachers would predict the future ability of the students who undertook it. They used the 'results' of this test to convince teachers that certain identified students in their class were gifted and would soon go on to display above-average improvements in their studies. Unbeknown to the teachers, the 'soon-to-blossom' students had been selected completely at random from among the class. When the students were assessed at the end of the school year, Rosenthal and Jacobson found that the preidentified students had indeed blossomed. The raised expectations of the teachers for the pupils had had the impact of actually raising the students' attainment.[131]

*

The fact that both animals and humans can pick up and respond to a second party's knowledge or biases, despite not sharing that knowledge or bias first-hand, motivates the need for so-called *double-blind* trials. In a clinical setting, both the subject and the researcher delivering the intervention are blinded to which treatment is being provided to which patient. This is straightforward enough for trials in which a medicine in pill form is being tested, but more complicated when trying to test some more invasive alternative therapies. Acupuncture, for example, involves the insertion of thin needles at specific points in the body. It is hard enough to replicate this experience for a patient receiving a sham treatment and almost impossible to deceive the individual administering that placebo.

Some clinical trials may go even further by blinding the patient, the researcher and some other party, like the research committee who oversee the study, in an attempt to avoid the inadvertent telegraphing of information to the researcher, who then passes it on to the patient. These trials are said to be *triple-blind*. In theory, this chain of inadvertent knowledge communication could go on indefinitely, needing higher and higher levels of blinding to better insulate the patient from the truth. In reality, practical concerns mean trials rarely go further than triple-blinding, and usually double-blinding is considered sufficient.

In the title of his book, *Of the Imagination, as a Cause and as a Cure of Disorders of the Body*, in which he first evidences the need for blinding to control for the impact of the self-fulfilling placebo effect, our tractor-debunking hero, John Haygarth, effectively hypothesises that the imagination could act not just as a cure, but also as a cause of ailments. Less well-known than the placebo effect is its malevolent cousin, the *nocebo effect*.

The nocebo effect is characterised by patients undergoing a procedure reporting a worsening of symptoms or other detrimental health impacts, with the outcome resulting not from the treatment itself, but from the subject's expectation that it might have a negative impact.

In his 2019 National Republican Congressional Committee speech, President Trump took the opportunity to brief strongly against renewable energy, suggesting that the noise from wind turbines could cause cancer. Living near wind turbines has been proposed to cause a much wider range of adverse health conditions, including headaches, dizziness, vertigo, nausea and heart palpitations. Despite numerous studies, no evidence that wind turbines cause any of these conditions has been found.[132] Instead, what has been uncovered is that the areas which show the highest levels of symptoms are those that have been exposed to the most negative information about the detrimental impact of wind farms.[133] Lab studies in which participants were played the noise of a wind farm have shown that the symptoms (or lack thereof) a subject reports depend strongly on the framing of the discourse surrounding the noise.[134] Those subjects who were not primed to expect a detrimental impact were unaffected by the noise, while those who were expecting one reported significant increases in the intensity and number of symptoms compared to pre-exposure levels.[135] This strongly suggests that many of the health claims that tend to fall under the banner of *wind-turbine syndrome* are most likely to be explained by a nocebo effect rather than any genuine causative link.

Conceptagion

The placebo and nocebo effects illustrate how strong the power of suggestion can be. If you expect to feel better, then, in many circumstances, you will. On the contrary, warning about a painful side effect might just make your experience of a treatment more uncomfortable. Here's an example you can play along with at home, which illustrates the power of suggestion (and here I'm sorry for what I am about to do to you).

It turns out that if you read about itchiness, or see someone scratching, you may well end up feeling itchy yourself and needing to scratch. Your itching display can then set someone else off. Although an epidemic of itching could, of course, be spread by itch-causing diseases, like scabies or chickenpox, for which there are physical vectors (the mite *Sarcoptes scabiei*, or the virus *Varicella zoster*, respectively), itching can be shared in a different way – through *social contagion*. I can afflict you with an itch from across the world while sitting in my living room, writing these words months or years before you read them (and I will confess that I caused myself to itch away for several hours while I researched and wrote the following paragraphs). Although it manifests itself physically, itch has the powerful conceptual contagion potential of an idea.

The autumn of 2001 was a time of extreme tension and heightened emotion in the United States. The September 11th attacks, the most deadly terrorist incidents the world had ever seen, cost the lives of almost 3000 people and injured thousands more. Just a week later, in the university town of Princeton, in the neighbouring state of New Jersey, five innocuous-looking letters were dropped into a mailbox. A few days later, these letters arrived in newsrooms in New York and Florida.

Without his glasses on, photojournalist Bob Stevens would have held the letter that was sent to the offices of the *National Enquirer* in Boca Raton, Florida, close to his face in order to read it. The message that came into focus likely read:

09-11-01. THIS IS NEXT. TAKE PENACILIN [*sic*] NOW. DEATH TO AMERICA. DEATH TO ISRAEL. ALLAH IS GREAT.

A few days after opening the letter, on a family trip to North Carolina, Stevens began to feel unwell, becoming feverish and flushed. On the drive back to Florida, his condition deteriorated and on the morning of 2 October, Stevens was admitted to hospital. He died just three days later, becoming the first of five fatalities and sixty-eight victims in total, of what would become the worst act of bioterrorism in US history. Unbeknown to Stevens, in addition to the block capitals that comprised the menacing message that covered the paper of the letter, it was also tainted with a fine powder that had been poured into the envelope meticulously by the sender – anthrax spores.

Against the background of these two momentous terrorist attacks, many Americans were left with a heightened sensitivity to any unusual happenings. Thousands of anthrax false alarms were raised across the United States that autumn, with 1200 of these reports in the state of Indiana alone.

On 4 October, the day that news of Stevens' anthrax infection in Florida was first brought to the world's attention, another unwanted medical problem was unfolding in Indiana. A third-grade student developed an itchy rash which quickly spread to other students in the class. The rash typically began on or near the face and spread to other exposed parts of the students' skin. The symptoms of the ailment seemed to dissipate when the children left school each day, only to return when they found themselves back in class, linking the illness quite decisively to the school setting. Strangely, none of the family members of affected children were reported to have been troubled by this seemingly contagious affliction. Although most parents chose not to voice their concerns openly, at the back of many of their minds was the distinct possibility that the outbreak was being caused by an unknown biological agent. Despite extensive investigations, no definitive explanation for the rash was ever discovered.

A month after the initial case, the eighteenth and final student to fall victim to the mystery itch eventually stopped scratching. Just as the enigmatic epidemic was burning itself out in Indiana, another rash of itching cases erupted at Marsteller Middle School, hundreds of miles away in Northern Virginia. This time about forty students and staff members across all the school's classes were afflicted. The school was forced to close while investigators searched for the cause.

Despite public-health crews in hazmat suits hunting for a possible environmental cause, nothing turned up. School officials were left scratching their heads – some literally. Upon reopening the school, after a thorough decontamination, hundreds more students became ill, some of whom had already been affected by the dreaded itch. As public interest in the story intensified, Debbie Files, a parent of one of the seventh-grade students who came down with the rash, voiced her concerns to the national newspapers: 'You know, the first thing you're thinking is that it's anthrax'. Despite these worries, heightened by the recent events, anthrax was quickly and conclusively ruled out.

Over the Christmas holidays, with students off school for a long period of time, reports of the rash in Marsteller students eventually dissipated. But on the return to school in January new and independent flare-ups were reported in schools in Oregon, Connecticut and Pennsylvania. Over the next few months, the school-related itch spread to all corners of the country. By the summer, outbreaks of the mysterious rash had been reported in over a hundred schools in twenty-seven states. There had even been some occurrences reported in neighbouring Canada.

Only very rarely were the rashes accompanied by symptoms like fever or vomiting, which would allow for a more definitive diagnosis. Occasionally, there was a positive test result for something like fifth disease, which can cause a rash and a fever, but typically, there were at most one or two such diagnosed cases in each outbreak. Some outbreaks were blamed by parents on the presence of mould in old, damp school buildings and others, more suspiciously, on misguided chemtrail conspiracy theories. The outbreak in one school was even suggested by some parents to be linked to dusty old maths textbooks. Despite these multiple purported but disputed causes, there seemed to be no common thread to link all the outbreaks together, other than the sudden collective onset of an itchy rash and the corresponding reduction of symptoms when the affected children were away from school.

After ruling out other possible causes, the conclusion that public-health officials investigating the majority of the outbreaks were forced to draw was that most of the cases of the rash were psycho-genic in origin – that the huge scale of the itching symptoms, which spread across the United States affecting so many students, was a form of *mass hysteria*. Such a diagnosis is rarely popular. Naïvely, it might seem that being told there is no concealed environmental threat would provide assurance to those afflicted. But for many, the diagnosis is perceived to be accompanied by the accusation of neuroticism or hypersensitivity. In particular, the word 'hysteria' can make sufferers and their families feel as if their genuine symp-toms are being diminished by know-it-all doctors insinuating it's all in their heads. Other more mistrusting individuals feel that their rational concerns are being glossed over as part of a wider-reaching coordinated conspiracy theory. Because psychosomatic illness is typ-ically a diagnosis of exclusion, the people charged with determining the cause of the outbreaks are also understandably reticent to make the call until every other possibility has been ruled out. No one wants to run the risk of being the official who missed the dangerous toxin or dismissed the contagious virus, especially where schoolchildren are concerned.

Nevertheless, despite its unpopularity, mass hysteria was the only tenable conclusion that explained the huge scale of the otherwise unconnected rash outbreaks. It is well known that skin is reactive to stress. Many people feel the hot flush of vasodilation creep up their necks when they are nervous. Eczema can be triggered or worsened by anxiety. Stress can elicit an outbreak of hives. The contagious power of itching against a background of students primed by the fear of anthrax poisoning and later by the national-media reports of the mysterious rash seems to have driven the epidemic.

Such outbreaks are variably called *mass psychogenic disorder, mass sociogenic illness, epidemic hysteria* or mass hysteria. These events are typically characterised by the rapid spread, between members of a social group, of symptoms that have no apparent known cause and for which no physical infectious agent can be identified. Some of the earliest recorded examples of mass psychogenic illness hail from Europe during the Middle Ages. From England down to Italy, a whole variety of different *dancing plagues* overtook schoolchildren, church congregations or even whole villages. Dancing could go on for weeks at a time, only halting when the victims fell down injured, exhausted or dead. One of the largest and most prominent dancing plagues took hold in July 1518 in the Alsace town of Strasbourg. Started by just a single individual dancing alone, the craze grew steadily, until a month later, over 400 people in the town had joined in. In a misguided attempt to help the victims 'dance away their mania', officials in the town hired musicians and erected an enormous stage for the merrymakers to help them burn off their energy. Unsurprisingly, the music drifting through the town and the sight of the revellers on the stage only attracted more people to the fray. At its height, fifteen people a day were reported to be dropping down dead, but still the dance went on. Until one day, it suddenly stopped with no obvious reason for the abrupt halt.

Another well-documented case of *hysterical contagion* occurred in the textile-manufacturing town of Spartanburg, South Carolina. Reports of a 'mysterious sickness' that had forced the shutdown of a textile factory in the town first came to the public's attention on the six o'clock news one hot Wednesday night in June 1962. At least eleven people, the report suggested, had been admitted to hospital with symptoms including nervousness, skin rashes, numbness, nausea and fainting. The agent assumed to be responsible for the rash was an insect that had escaped from a shipment of cloth, although exactly what type of insect no one could quite be sure.

Over the coming days, more and more of the factory's workers came down with the unexplained illness. By the weekend, the patient list numbered sixty-two. An entomologist was brought in to try to isolate the culpable insect and the plant was fumigated. Despite finding several species of insect and mites in the plant, some of which could have delivered mild bites to the workers, none was isolated that alone could have caused symptoms severe enough to require hospitalisation. As one exterminator who inspected the factory put it, 'Whatever has been here, ain't here now'.

In fact, sociologists who interviewed the workers at the plant found that the main predictors of who became ill were background anxiety levels, the amount of overtime worked and responsibility for earning a significant proportion of their household's income.[136] Clusters of individuals affected at similar times were typically found to belong to tight-knit cliques. In short, the sociologists had identified the classical conditions (background anxiety and close social ties) for hysterical contagion. They concluded that, while an insect bite may have been the trigger for some patients, causing an itchy rash, the remainder of the symptoms were likely psychogenic in nature. The fainting and nausea which had caused the plant to be shut down were likely a result of a social epidemic – an illness spread by the infectious power of emotion, rather than a physical vector.

*

The positive-feedback loop we met at the start of the chapter and which caused Covid cases to rise exponentially was mediated by a physical agent – the SARS-CoV-2 virus. However, just because an illness is spread by an idea or an emotion, rather than a viral or bacterial vector, it doesn't make that illness any less real for the communities or individuals affected. The same mathematics that we use to describe the explosive onset of an infectious disease can be used to describe the viral outbreak of an idea – a conceptagion. Scientists have suggested that a hugely diverse range of social phenomena – from generosity to violence and from kindness to unemployment – may be socially contagious. Some scientists have even come full circle by suggesting that diseases like obesity and insomnia, which are typically considered to be non-communicable disorders, may have a strong social component that allows them to spread like a contagious disease. Whether teen pregnancy, for example, is genuinely socially contagious, as some scientists claim, is still hotly debated.

What is clear, however, is that a concept-mediated contagion can be more difficult to wipe out than one for which the vector of spread is more tangible. Even ideas that have lain dormant for hundreds of years can take off – amplified from person to person through a positive feedback loop. Metaphorically, they are the residual vials of smallpox that are held in the lab ready to be unleashed, by the clichéd rogue scientist, on an unsuspecting susceptible population.

Underestimating an idea's potency, its longevity and its ability to enthral can lead us to misjudge or misunderstand how a situation will play out. One only has to look at the pervasive spread of disinformation throughout the Covid-19 pandemic to see the damage that dangerously incorrect ideas – overstating dangers of safe and effective vaccines, underplaying the risks associated with contracting Covid and upselling the effectiveness of unproven treatments – can do. The viral spread of these myths through social media means they can reach far and wide in virtually no time and are, consequently, extremely difficult to counter. We underestimate the snowballing of these pervasive myths at our peril.

It's no exaggeration to say that people have died as a result of these sorts of falsehoods. Indeed, everyone stands to lose out through misinformation, as preventable diseases like measles and polio, which could be eliminated with high enough vaccination rates, are allowed to gain a foothold.

We must learn to fight fire with fire – or rather snowballs with snowballs – making the truth as attractive and transmissible as the fake news that would sabotage it. It may seem harder to send evidence about the effectiveness of vaccines viral than it is to spread a sensational anecdote which disparages them, but it is in our gift. When we click 'like' on social media, or even just when conversing with our friends and neighbours, it's vital that we share trusted sources of information and challenge fake news where we find it.

In a political context, it's not possible to put the spread of an idea to the sword like it's a monster in an action movie. An ideology can't be decapitated by executing the movement's figureheads, as the West has rediscovered to its cost time and time again – most recently when attempting to defeat al-Qaeda and ISIS; hydra-like, another figurehead simply grows to prominence to replace the one that was removed. Instead, a new and more attractive ideology must be cultured from the grassroots, displacing and disrupting the status quo. Temporarily imposing a new set of values on a population doesn't work, as we have seen with tragic consequences with the resurgence of the Taliban in Afghanistan.

Indeed, such attempts can rebound with far-reaching consequences. As a case in point, it was, in part, the continued attempts to suppress democracy that led to the domino effect that saw country after country topple its dictators in the so-called 'Arab Spring' of 2011 – the success of each country emboldening the actions of the next, in a positive feedback loop. As we will examine in the next chapter, the attempt to contain an idea or a movement all too often ends up diverting attention towards the subject and boosting it to greater prominence.

8

CATCHING BOOMERANGS

Since they first came to the world's attention over a decade ago through their reality TV show, the Kardashians have built an empire. Each of the Kardashian-Jenner sisters fronts a multi-million-dollar personal brand and together they maintain consistently high levels of coverage in both traditional and social media. They are idolised by millions of women and girls as the perceived embodiments of perfection. They sell their acolytes an aspirational dream of health, wealth and beauty.

Perhaps one of the more overlooked of the sisters, although still incredibly popular in her own right, is Khloe. She has built her brand around body positivity and fitness. Her spin-off reality TV show, *Revenge Body*, is all about empowering 'ordinary' Americans to deal with their own self-image insecurities. She has a denim clothing brand whose tagline is 'Representing body acceptance' and her workout videos help to push a line of fitness products. All her business efforts are bolstered by a carefully curated social media campaign in which she flaunts images of her body as a demonstration of what her followers might aspire to achieve if they buy in.

Indeed, near-impossible body images, used to sell cosmetics, clothing and even appetite-suppressant lollipops, are a mainstay of the Kardashian brand. In recent years, eagle-eyed followers have gathered and drawn attention to a growing portfolio of evidence suggesting that many of these images are heavily doctored. Yet despite increasing numbers of fans cottoning on to the fact that the Kardashians are selling an altered brand of aspirational reality, their marketing power remains undiminished. It's almost as if the propagation of these apparitions of perfection is enough for their fans; irrespective of whether they represent heavily curated, unachievable perfection created through digital manipulation.

Given the importance to the Kardashian brand of perfectly manicured images, it was surprising to read reports of an unedited photo of Khloe Kardashian appearing online over the Easter weekend in early April 2021. Although not perpetuating the bronzed, toned, impossibly-thin-waist look closely associated with the Kardashians, very few people would have suggested that the image – showing Khloe standing by the pool wearing a leopard-print string bikini – was unflattering. While it seemed a marked change of tack from the sisters' typical marketing ploys, this undoctored image of Khloe, warts and all, seemed to be on-brand with her body-positive message.

However, it soon transpired that this was not a marketing ploy, but a mistake. Within hours of releasing the photo on to the internet the Kardashian legal team had snapped into overdrive in an attempt to claw it back. Reposted copies of the photo were forced down under the threat of legal action. Twitter accounts that had reposted it were temporarily suspended. The Kardashian camp issued a statement suggesting the image had been 'posted to social media without permission and by mistake by an assistant'. This reaction turned out to be a critical mistake.

On 6 April, just hours after the statement was issued, celebrity news and gossip site Pagesix.com ran a story about the leak in which they republished the photo. Over the next twenty-four hours, Google searches for 'Khloe Kardashian' jumped to twenty-five times their previous level. In the following couple of days, media outlets across the globe, including the *Daily Mail* – the most visited English-language newspaper website in the world – were running copies of the photo as a news item. Google searches for the influencer were at fifty times the pre-statement levels. People who would never ordinarily have come across the photo were having it thrust into their newsfeeds. This is how I first came across the story. Others were actively searching to find it. Ordinarily, I, like many others around the world, would never have seen it or even known of its existence.

Of course, as genuine as the attempts to censor the image may have appeared, it's impossible to rule out yet another clever marketing ruse from team Kardashian. In a world in which (almost) all publicity is good publicity, it certainly can't be ruled out that the spike in internet-search activity and the prominent newspaper headlines generated by the 'leak' were the deliberate result of a successful marketing stunt by someone in the camp who was only too aware of the *Streisand effect*.

The Streisand effect is the name given to the increased publicity attracted to a piece of information by an attempt to remove or censor it. In 2002, environmentalists Kenneth and Gabrielle Adelman undertook the epic task of taking photographs of all of California's coastline in order to document coastal erosion. Once completed, they made the 12,200 photographs publicly available on their website. One of them happened to include Barbara Streisand's Malibu mansion and Streisand did not take kindly to a photo of her residence being freely available on the internet, so she decided to sue the Adelmans. At the time she filed the suit, the photo had been downloaded six times – including twice by Streisand's lawyers and once by her neighbour. In the month following the suit, nearly half a million people visited the Adelmans' otherwise unremarkable website. In the event, Streisand lost her suit and was forced to pay $155,000 in legal fees.

As super-injunction litigants and book-banners have discovered, there are few better ways to pique interest in something than by prohibiting it. Apple fell foul of the Streisand effect when they filed a legal complaint against a book which they claimed disclosed 'a multitude of business secrets'. After the complaint came to light, *App Store Confidential*, by former head of the Apple App Store in Germany, Austria and Switzerland, Tom Sadowski, quickly sold out its first print run of 4000 copies, rising to number two on the Amazon best-seller list in Germany. Despite reviews describing the revelations in the book as 'bland and obvious', Apple's litigation left the public wanting to find out exactly what it was the tech giant were so worried about.

In a political context, a 2019 study found that while the imprisonment of online critics of the Saudi Arabian government did tend to deter them from further dissent, it did little to dissuade others.[137] In fact, social media followers of imprisoned dissidents were found to be energised by the incarcerations, leading to stronger and louder calls for political reform and regime change.

The Church of Scientology has also fallen victim to the Streisand effect. When a video of Tom Cruise ranting incoherently about scientology was leaked on the internet in January 2008, the Church quickly filed a copyright claim in an attempt to get it taken down. Not only did this attempt at suppression significantly increase views and serve to spawn copies of the video, it also resulted in the creation of a committed anti-Scientology campaign. Anonymous, the international hacktivist collective, interpreted the Church's actions as attempted internet censorship and declared their aim to expel the Church from the internet. The vigilante group's direct actions included distributed denial of service attacks, which rendered the Church of Scientology's official website temporarily inaccessible, and the leaking of private documents, allegedly stolen from Scientology sources. Without the Church's censorship attempt, it seems extremely likely that interest in the video would have fizzled out quickly and it's almost certain that they would not have drawn fire from Anonymous.

This chapter is all about boomerangs and how to spot them. Boomerangs are actions that, although potentially well intentioned, can dramatically veer off course without due consideration for their potential consequences. They are the proposed 'solutions' which make a problem worse not better, or the predictions which change the future and negate themselves. In the most extreme cases, the boomerang can complete a full 180 and come back to hit you in the head, sometimes over and over again. There are instances in which you might not even know you've thrown a boomerang until it's too late, but there are telltale signs to look out for. By recognising their characteristic signatures, we'll learn to catch or divert these potential own goals or, better still, with a little foresight, learn never to pick up a boomerang in the first place.

The Streisand effect is just such an example. It relies on the psychological phenomenon of *reactance*, otherwise known as *the boomerang effect* – a motivational reaction that occurs when people are under the impression that their choices are being limited, their freedoms curtailed or that they are being persuaded to do something they don't want to do. Reactance can make it harder to persuade someone of your point of view or even cause them to adopt the opposite view to the one you hoped to impress upon them. It is reactance that I rely on when I tell my kids, 'I bet you can't get upstairs and ready for bed in five minutes'. Reactance is at the heart of *reverse psychology*.

As part of a pair of studies into reactance, students were subjected to health messages[138] – one encouraging flossing and another discouraging drinking. The messages warned about the negative health consequences associated with not flossing or with binge drinking, and suggested how to change behaviour to avoid those impacts. For some students, the behavioural portion of the message comprised a strong direction, mandating participants to either take up or continue flossing, or alternatively to completely avoid binge drinking. For others, the message was milder, merely encouraging or suggesting the same behavioural changes. Students subjected to the stronger messages were found to exhibit more traits associated with reactance (perceiving a more significant threat to their freedom, having more negative thoughts and experiencing stronger feelings of anger) than those subjected to the milder messaging. The degree to which a subject perceives that their freedom is under threat is a strong indicator of the strength of reactance: the most obvious way for a subject to neutralise the threat and reassert their freedoms is to embrace the proscribed attitude or engage in the discouraged behaviour. In short, the study suggested that the students subjected to the harsher messaging, which engendered more reactance, were subsequently more likely to binge drink and less likely to floss.

Another study looked into the impact of newly placed 'No Diving' signs positioned at the shallow end of school swimming pools.[139] Not only were high-school students with a history of diving in at the shallow end more likely to notice the signs, but they reported being more likely to repeat the proscribed plunges when the signs were present than when they weren't.

The students' behaviour in the above studies is indicative of the causes of the failures of many public-health campaigns. When studying the impact of the guidance 'Warning by HM Government. Smoking can damage your health' on cigarette advertisements in the UK, researchers found the messaging actually increased the subjects' desire to smoke.[140] Similarly, US researchers found that exposing adolescents to cigarette warning labels, such as 'SURGEON GENERAL'S WARNING: Quitting Smoking Now Greatly Reduces Serious Risks To Your Health' led to a significant increase in smoking in comparison to teenagers not exposed to the messaging.[141] Some public-health psychologists have suggested that the cost of engendering oppositional behaviours and attitudes when conducting public-health campaigns around subjects like the dangers of alcohol abuse or smoking might outweigh the meagre benefits in consumer knowledge gained through such messaging.[142]

Perverse incentives

Another class of boomerangs, which doesn't rely generally on reactance, falls under the umbrella of *perverse incentives*. These are incentives offered with the aim of achieving a particular goal which, unexpectedly, end up bringing about the opposite of the intended

impact. An alternative and commonly used name for the phenomenon is the *cobra effect*. The name derives from a story which goes
back to the era of the British Raj in India.

Bureaucrats in Delhi became concerned about the number of
venomous cobras in the city. In order to resolve the problem, they
placed a bounty on each cobra's head. Every cobra carcass brought
to officials could be exchanged for a monetary reward. Soon after
the announcement, thousands upon thousands of dead cobras began
to pour in. The policy appeared to be a great success. However, not
everything was quite as expected.

Rather than doing the hard work of going out and catching cobras,
some entrepreneurial individuals had taken the initiative to set up
lucrative cobra-breeding programmes to cash in. While the British
continued to buy up cobra corpses, the number of cobras on Delhi's
streets remained low. But as soon as the British got wind of the
breeding ruse, they withdrew the bounty in order to remove the
incentive for these rogue cobra farmers. Of course, with no viable
income stream, the temporary serpent-breeders couldn't afford to
support their now-worthless animals, so, instead, decided to release
the cobras, in huge numbers. As a result, the story goes, the itinerant
cobra problem became worse than it had ever been.

You would have thought the occupying British would have learned
their lesson, but in 2002, they made a similar mistake in Afghanistan.
Despite it constituting a lucrative source of income, in 2000, the
Taliban's then leader, Mullah Omar, had declared opium un-Islamic.
Wary of the impact of disobeying the Taliban's orders, Afghan farmers
reduced their profitable poppy cultivation by nearly 90 per cent
between 2000 and 2001. After the overthrow of the Taliban, as a
result of the US-led invasion in late 2001, Afghan farmers quickly

went back to planting the poppies whose sap is the key ingredient in both morphine and heroin. While the US military switched their attention to hunting for Al-Qaeda targets, including Osama Bin Laden, President Bush called upon his NATO allies to help control the resurgent opium-production problem in the country. The British quickly stepped up to answer their ally's call for help.

Lacking the powers of coercion wielded by the former Taliban leaders, the British decided to tackle the problem with a carrot rather than a stick. Afghan farmers were to be offered $700 for every acre of poppy crops they destroyed. This was an absolute fortune for many impoverished poppy farmers, who quickly signed up to the scheme. Tens of thousands of acres of poppies were destroyed as part of the programme. Unfortunately, far more were planted to compensate for the fields that were lost. Many farmers harvested the opium sap before destroying the poppies, ensuring they would reap a double payday. By the time the British eventually pulled out of Afghanistan, four times as much land was being used to cultivate poppies than before the introduction of their perverse incentive scheme.

These sorts of mistakes are, of course, not limited to the hapless British occupying foreign nations. Other governments have made similar errors in their own back yards. In the 1860s, the US government, for example, was employing two railroad companies to build the Transcontinental Railroad. The Central Pacific Railroad company was to build eastwards from Sacramento, while the Union Pacific Railroad company was to build west from Omaha, with the aim that the tracks would meet somewhere in the middle. The federal government made the mistake of remunerating the companies by mile of track laid. This incentivised both outfits to build overlong tracks which went by circuitous routes, which they duly did. The two companies were also aware that every mile the other company laid in one direction would reduce the amount of mileage they could cover in the other direction. This encouraged fast, but often shoddy, workmanship in the race to lay the most track. As the two tracks began to approach each other in Utah and the per-mile cash cow looked to be drying up, the two companies tacitly agreed to carry on building their tracks in parallel right past each other. Even after the two tracks were eventually made to meet, it took years for the rerouting to be completed and the repairs to be carried out in order to make the track serviceable.

Goodhart's law

Naïvely constructed incentive systems are a classic way in which undesirable outcomes can be inadvertently encouraged. In particular, the setting of targets which don't tackle the root cause of a problem, but instead address only a proxy for the underlying issue, can lead to a scheme backfiring.

For decades, successive Colombian governments had battled the left-wing Revolutionary Armed Forces of Colombia (FARC) guerrilla group. Founded during the cold war on Marxist–Leninist principles, with the aim of promoting anti-imperialism and peasants' rights, FARC funded its military activities through a combination of ransoms, illegal mining, extortion and the production and distribution of illegal drugs.

In the early 2000s, after over thirty years of ever-intensifying civil war, the Colombian military decided to step up their assault on FARC. Their plan was to render the guerrillas' operations untenable by simply wiping out its adherents. To that end, they devised league tables and a reward system. Senior army officers ranked military units by the numbers of enemy fighters who were killed, captured or surrendered, in that order. Killings became the top priority. Capturing enemy combatants alive was disincentivised.

In 2007, the army's seventh division released a campaign plan for the following three years. The document states unambiguously that the division's units will be assessed primarily by the number of deaths of enemy soldiers reported. The reduction in the number of terrorist incidents, which might have made a more suitable target for the army's overall strategic aims, was listed only as a secondary objective. At an individual level, commanders and soldiers who reported high numbers of kills were rewarded with money, time off and distinguished service awards. A leaked army document called 'Policies of Gen. Mario Montoya', chief of the army from 2006 to 2008, goes as far as emphasising that 'Kills are not the most important issue, they are the only issue'.

As a result, many army units, finding it difficult to hunt down and eliminate the elusive and well-trained rebel soldiers, resorted to murdering civilians in order to increase their kill count. Young men, often from poor families, were lured from their homes with the promise of work and murdered in cold blood. The problem was so widespread and pervasive that between 2002 and 2008 the Colombian military were found to have killed over 6000 of their own citizens – the very people they were supposedly being deployed to protect. The indiscriminate target of killing rather than taking captive or eliciting the surrender of these innocents, who came to be known as the 'false positives', allowed the soldiers to pass off the dead as left-wing rebels in a way which would have been much more difficult if the target had been to capture the combatants alive. Dead men, as they say, tell no tales.

The perverse incentive offered to the Colombian army is a classic example of *Goodhart's law*, which states that, 'when a measure becomes a target, it ceases to be a good measure'. Before the target was introduced, and consequently, before the murder of civilians was incentivised, most killings at the army's hands would presumably have been of FARC rebels. All other things being equal, this metric may have provided a good measure of how well or poorly a unit was functioning. As soon as 'lives taken' became a target, irrespective of whose lives they were, there were clear incentives to game the system and the raw number of killings ceased to be a good metric of a unit's effectiveness.

We live in a world of league tables. We rank everything from the schools and universities our children attend to the hospitals where we expect our most vulnerable to be cared for. League tables can be based on a range of metrics. If those metrics are chosen appropriately, the rankings can give useful insight into the quality of an organisation, allowing us to reward those near the top and support those lagging behind. However, unless the performance indicators are chosen extremely carefully, we may find institutions optimising their performance towards the metrics, rather than the underlying goals for which those metrics are a proxy.

Schools are a case in point. What we really want to assess in schools is the quality of teaching and learning opportunities. This should be a fundamental measure of a school's academic qualities. However, because assessment of these qualities is labour-intensive, schools are typically assessed instead on student-test outcomes, a rough proxy for teaching quality. The rationale behind the choice is clear: the better the teaching, the better the student-test outcomes should be. But, of course, there is a loophole. Student-test outcomes can be improved without improving the teaching.

This is exactly the issue that led thirty-four teachers and principals in the Atlanta school system to be indicted for racketeering in 2015. Nationwide education reforms saw states compete with each other for federal funding linked directly to high and improved test scores. This put pressure on state education administrators, who compelled district school superintendents to improve scores. In turn, the superintendents coerced schools and individual teachers into manipulating test scores by offering them financial incentives or threatening punishments. Impelled by these pressures, a large group of educators conspired to fix the results of standardised tests in order to bump their schools up the league tables. Eleven teachers were eventually convicted as part of one of the worst occurrences of test cheating the United States has ever seen. As the old proverb goes, 'Be careful what you reward, for you will surely get it'.

The maxim holds not just in schools, but in any area in which targets are regularly used to rank competing institutions. Unfortunately, one arena in which such target-based contradistinction is prevalent, and consequently in which manipulating metrics can facilitate the climbing of league tables, is healthcare. If the targets are not chosen with the utmost care, the consequences can be disastrous, as Vietnam veteran Walter Savage discovered to his detriment.

After a nasty fall, eighty-one-year-old Walter hobbled out of the cold, December, Oregon night into the brightly lit emergency department of the Roseburg Veterans Affairs hospital. Suffering from dehydration, malnutrition and broken ribs as a result of his fall, Walter was examined by a doctor who decided he needed to be admitted. However, despite there being plenty of beds available that evening, the hospital administrators judged that Walter was not sick enough to be admitted. After a punishing nine-hour wait, while doctors argued with administrators over the seriousness of his condition, Walter was eventually discharged to make his own way back home, alone. The next day he came back to the hospital. Again, he was made to wait hour after hour to be admitted. Eventually, the attending physician, refusing to take no for an answer, admitted Walter. After being seen briefly by the doctor, the hospital administration ensured that he was shipped off to a nursing home less than twenty-four hours later.

The reason for Walter's initial discharge and subsequent transferral wasn't that he wasn't unwell, or that he didn't need urgent medical treatment. If anything, it was the opposite. Hospital administrators seemed concerned that Walter's case might blot their copybook. Two years previously, the hospital had begun a policy of prioritising the lowest risk and simplest-to-treat patients in order to boost their scores in the Department for Veterans Affairs' gradings system. Veterans were moved on as quickly as possible, ostensibly because the hospital was 'too small' to treat the most serious patients. One doctor working at the hospital claimed, 'It's a numbers game. The leadership has figured out the hospital can actually do better by seeing less [sic] patients.' He wasn't wrong. Under this policy Roseburg climbed the rankings as one of the 'Fastest Improved Hospitals in Healthcare Quality'. No doubt the direct link between the metrics and the leaderships' bonuses didn't diminish their enthusiasm for climbing the ladder.

This wasn't the first time the veterans' hospital system had been hit by a ratings-related scandal either. In the late nineties, the Department of Veterans' Affairs added surgical complications to its list of hospital metrics. The more surgical complications recorded, the lower the hospital would be ranked. The policy seemed to work. Remarkably, between 1997 and 2007, reports of complications fell by almost half. In reality, however, compelled by hospital administrators, many doctors simply stopped taking on the most complicated and high-risk operations. One doctor who worked in the system at the time admitted that patients lost their lives because of the policy. In situations for which the metric against which quality is judged is gameable, the provision of strong incentives can actually lead to diminished, rather than improved, outcomes.

Sheep learning

It's very easy to attribute the exploitation of loopholes in the rules to our human instincts. Galvanised by pride, or greed, vainglory or embarrassment, it's easy for us to imagine the motivations of those who are so driven to game the system. However, it turns out that algorithms, lists of instructions for performing tasks, which we often consider to be immune to the subtleties of human artifice, can be just as prone to taking these shortcuts as we are.

Recent advances in *deep learning* are the driver of much of the excitement surrounding the field of artificial intelligence. Deep-learning algorithms are designed to acquire competence in an activity or task by roughly mimicking the biological processes underlying the way in which the human brain learns. For image classification, an algorithm might be given a set of pictures, some of which contain the object being searched for and others which don't. For example, when attempting to spot grazing animals in an image, the deep-learning method might be trained on some 'positive' images which contain sheep, cows, horses, etc. and other negative images which contain cars, or fire hydrants or traffic lights or a whole array of other real-world scenes.

For this *training data set*, the algorithm will be told which is which, so it can analyse and search for features in the image which are good predictors of the object to be identified. Ideally, it will hone in on the features which best characterise the object: its shape, colour, contrast or perhaps some other less obvious traits. The choice of relevant features is left to the algorithm. Images with cows in them might teach the algorithm to detect rotund objects with contrasting black and white patterns, for example. In part, the freedom to determine the important features in the image is what makes deep learning such a powerful technique. The ability to choose any identifier means that algorithms can discriminate between images using metrics it hasn't been taught by human trainers and which may not be obvious to, or even be detectable by, the human eye.

Before it is unleashed on the world, the algorithm would typically be validated on a *testing data set*. This second set of images, for which the correct classification is also known to the algorithm's human creator but not to the algorithm itself, allows researchers to determine how well or poorly the algorithm has 'learned' to do its intended job.

The field has recently achieved great success, with the development of algorithms able to beat Grand Masters in the ancient Chinese game of Go[143] or take professional poker players to the cleaners.[144] The tools of deep learning are already being exploited commercially by companies like Facebook to automatically tag people in uploaded photos, and Google to translate text between over a hundred different languages. In the medical arena, deep-learning algorithms have been able to detect cancer from X-rays with an accuracy that rivals that of human radiologists.[145] The potential applications are limitless.

However, there have been a number of embarrassing failures for the emerging technology as a result of algorithms taking shortcuts to achieve the human-imposed target. Image auto-captioning, part of the field of computer vision, is theoretically an incredibly useful tool enabling search engines to present the most relevant image search results and improving accessibility. Unfortunately, because no rules are laid down to guide the algorithm about how it should go about associating the input data with its classification, it is all too easy for artificial-intelligence algorithms to take shortcuts. One algorithm, trained to identify grazing animals, learned to find them quite accurately by picking up on the large amounts of green grass in the background of the training images, rather than finding the more difficult-to-identify animals themselves.[146] The end result was that many empty landscapes ended up being misclassified with captions like 'grazing sheep' or 'a herd of cows'. Mildly annoying, perhaps, if a mislabelled image turns up in your search, but perhaps not that consequential. The same cannot be said for other areas in which these machine-learning algorithms are taking shortcuts.

One of the great hopes for such artificial-intelligence algorithms in healthcare settings is that they might relieve some of the burden on overloaded staff. In particular, radiologists, whose job it is to look at and interpret many different types of medical images, from X-rays to CT scans, stand to benefit significantly from algorithmic assistance. Of course, in order to reduce their burden, we need to be able to rely on the outputs of the algorithms' decisions to a similar degree to which we trust the human-made classifications. Unfortunately, it's sometimes difficult to be completely convinced that an algorithm with unknown inner workings will always perform as expected. Indeed, a large part of what unnerves many people about the potential use of artificial intelligence in applications from driverless cars to clinical diagnostics, where lives are potentially at stake, is that the internal logistics of such algorithms are quite opaque. Many machine-learning processes are so called *black boxes* for which not even the designer of the algorithm understands what's going on under the hood. They can appear to function correctly under one set of conditions, but then malfunction catastrophically in other similar circumstances, often without leaving any clues to suggest why they failed.

Even more troubling is that it may not even be possible to predict on which classes of problems or individual scenarios the algorithms might break down until those situations are encountered for the first time. For example, the addition of background noise to human speech has fooled speech-recognition systems into hearing phantom words or phrases. The simple addition of a few small stickers has tricked computer-vision algorithms for self-driving cars into misreading warnings that should have told the driver to 'Stop' as speed-limit signs.[147] Released on the real world, the consequences could be disastrous. Since the internal operations of some deep-learning algorithms are essentially black boxes, there is nothing to stop them taking shortcuts to achieve their aims – shortcuts which allow them to perform their tasks quickly and reliably in the settings of the data on which they are trained, but which perhaps bear no resemblance to the true path required to tackle the general class of problems for which they are designed.

One deep-learning algorithm, designed to diagnose pneumonia, fell into precisely that trap.[148] After having spent time learning from a training set of lung X-rays, when pitted against the testing data set, the computer-vision algorithm was found to be doing a good job at highlighting lungs with pneumonia and ones without. It wasn't perfect by any means, but passable in comparison to the human benchmark. Before putting the algorithm to work in a real healthcare setting, the developers carried out some simple tests to see if they could better understand and improve the workings of their method – and it was a good job they did. By inspecting the areas of the X-ray image that the algorithm was focusing on primarily when making its assessments, they found, bizarrely, that it was assigning most weight to a region around the shoulder of each patient, almost completely ignoring the lungs themselves. So how was it able to do even a passable job at identifying pneumonia without examining the lungs, whose inflammation is the very definition of the condition?

Often, when imaging roughly symmetric body parts, such as the lungs, radiographers will place, somewhere in the field of view, a metal disc with the shape of an L or an R outlined, to allow the left side of the image to be distinguished from the right. The algorithm's seemingly miraculous diagnostic ability was actually a result of the fact that the two hospitals from which the testing and training data were drawn, had quite different rates of pneumonia. As well as having quite different prevalence of pneumonia, the two hospitals also used different left/right marker tokens, which the algorithm was able to tell apart. Instead of searching the images for indications of lung inflammation, the algorithm instead detected the outlines of the different tokens (usually in one corner of the X-ray, close to the patient's shoulder) allowing it to identify the hospital and then to use what it had learned about the hospitals' pneumonia prevalence to make its classification.[149]

The great strength of machine learning – the freedom of the algorithm to find the most appropriate way to interpret the data to meet the stated target – is also the significant weakness. Such an algorithm is allowed to take shortcuts or use overly complicated metrics that may have no genuine relevance to the desired outcome, meaning that they don't generalise well. What the algorithms are doing under the hood is trying to build a model which relates the input data to the desired output. Here, there is an argument for saying simple models should be preferred over more complicated ones.

Unfit for purpose

More broadly, when trying to acclimatise a model to training data, the algorithm can choose to pay as much or as little attention to the features of those data as we choose. Too little attention and the model may miss the important underlying trends; too much and it ends up misinterpreting random fluctuations or errors as meaningful patterns. These problems are known as underfitting and overfitting, respectively.

For some training data sets the most appropriate model is clear. For example, if I ask you to give me the next number in the following sequence 3, 5, 7, 9, you might reasonably expect the answer to be 11. Looking at the first few terms a simple pattern emerges. To get from one term to the next we add 2. Implicitly, the model we have fitted to the data is linear – increasing by a fixed amount each time we go to the next term. The formula $2 \times n + 1$ tells us the value of the nth entry of the sequence. The expression tells us that the first term (n takes the value 1) in the sequence has the value $2 \times 1 + 1$, which gives 3. The second term (n takes the value 2) can be evaluated as $2 \times 2 + 1$, giving 5 and so on. This leads us to predict the fifth term in the sequence should be $2 \times 5 + 1$, which gives us our expected value of 11. This is a completely reasonable assumption and the simplest model that fits the data.

However, another more complicated model of the form $13 \times n^4 - 23 \times n^3 + 17.5 \times n^2 - 5 \times n + 0.5$ does just as good a job on the training set. Try setting n equal to 1 and the formula $13 - 23 + 17.5 -5 +0.5$ gives 3 – the first term in the sequence. Try setting n equal to 2 and we get 5, the second term in the sequence. And so on. However, when we plug in 5 to find the unspecified fifth term in the sequence, we get the answer 23, not 11 - a very different answer to the one we might intuitively have expected. The answer 23 is not wrong, but in some senses it is overcomplicating the data. We are never likely to dream it up for ourselves, because it relies on a model which seems overly complicated. The curve representing the simple linear model is given in the left-hand panel of Figure 8-1, whereas the curve capturing the more complicated model is given in the right-hand panel. While the linear curve fits a simple straight line through the first four data entries, the complicated model wiggles about in an unexpected way between the points.

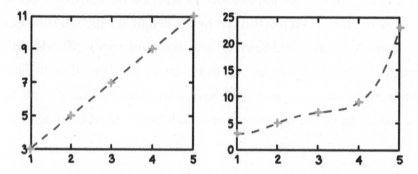

Figure 8-1: The linear model (in the left panel) fits a straight line through the first four data points to predict the fifth to be 11. The more complicated model (in the right panel) fits a nonlinear curve through the first four data points to give a very different prediction for the fifth. Neither model is 'correct' in any objective sense, but the linear model may be preferred because of its simplicity.

A general rule of modelling is known as *Occam's razor*. In this context, the rule can be interpreted as the maxim 'use the simplest model possible that explains the data'. The linear model in the left-hand panel of Figure 8-1 does exactly that, while the right-hand panel clearly does not. While applying Occam's razor sounds simple in principle, in real-world data sets it's not always obvious what is the true underlying trend and what is noise in the data. Simple models might try to ignore the noise at the cost of also missing the signal and, consequently, make a poor prediction. Conversely, complex models might give too little weight to the signal in the training data by paying too much attention to the noise.

When we talk about the simplicity or complexity of a model, we often refer to the number of *parameters* or *degrees of freedom* the model has. The parameters are just the quantities we need to extract from the data in order to characterise the model. For example, a linear model has two parameters: the slope of the line and how high up or low down it is (we often characterise this last by the height at which the line hits the vertical (y) axis and refer to it as the *y-intercept*). Models with more parameters provide us with more flexibility when fitting to data. However, the more parameters we have, the harder it is to be sure about the correct value for any one of them with only limited training data.

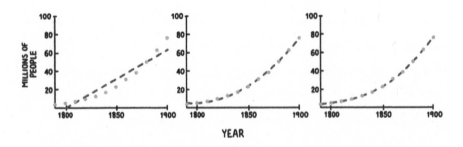

Figure 8-2: Different model fits (dashed lines) to US population data (dots) from 1790 to 1900. The linear model in the left panel underfits the training data. The three-parameter model in the middle panel does a good job of fitting to the training data, missing only a few data points by a small margin. The seven-parameter model on the right does an even better job with the dashed line going through almost all the training data.

In Figure 8-2, I've plotted the US population size taken from census data every ten years between 1790 and 1900. I've fitted three different models of increasing complexity against the training population data with a view to predicting how the population evolved in the twentieth century: the 'basic' linear model shown in the left panel, with two parameters; the 'improved' model in the middle panel, with three parameters; and the 'complex' model displayed in the right panel, with seven parameters.

The basic model does a poor job of mimicking the training data. Although it describes the fact that the population is growing, it does not have enough freedom to capture the fact that the rate of growth is itself increasing. The improved model with one extra parameter fits most of the data points fairly well, only narrowly missing one or two. The complex model has enough parameters that it does an excellent job of fitting closely to almost all the data points.

When we look at the predictions made by each model in Figure 8-3, we see that, as expected, the basic linear model fails to predict the increasingly rapid growth of the population after 1900. The improved model does a good job of tracking the population in the post-1900 testing data set. Perhaps surprisingly, the complex model, which fitted the training data so well, does the worst job of all, predicting rapid population decline followed by extinction by the mid-1900s. Although the complex model fitted the training data better than the improved model, by doing so it paid too much attention to the variation in the data at the expense of the general underlying trend. More complicated models are not always better.

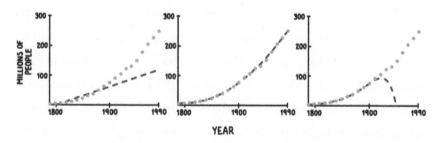

Figure 8-3 The basic linear model (left panel) does a poor job of predicting the testing data. The overparameterised complex model (right panel) does even worse. The improved three-parameter model (middle panel) does a good job of predicting the US population well into the twentieth century.

When fitting to real-world data, we can never expect to match all the data points perfectly. Indeed, we shouldn't hope to do this. Instead, we should decide an error tolerance that we are happy to accept. Occam's razor – keeping the model as simple as possible – suggests that the improved three-parameter model was the simplest one that fit the training data set to within a reasonable tolerance. Hence, it should have been the model we selected. With only twelve data points to determine seven parameters, the complex model could be described as overkill. We shouldn't be surprised when a complex model with many parameters gives a good fit to training data. We should only be impressed when the predictions of that model have been borne out by reality. This is exactly what the famous mathematician, John von Neumann, was warning about when he said, 'With four parameters I can fit an elephant, and with five I can make him wiggle his trunk.'

Self-defeating prophecies

Predicting population growth is fraught with danger. In 1968 in his book *The Population Bomb*, American biologist Paul Ehrlich, musing on the potential food shortages that might result from a rapidly growing population[150] wrote, 'The battle to feed all of humanity is over. In the 1970s hundreds of millions of people will starve to death in spite of any crash programs embarked upon now. At this late date nothing can prevent a substantial increase in the world death rate.' He continued by quantifying his predictions: 'The death rate will increase until at least 100–200 million people per year will be starving to death during the next ten years'.

Even the most cursory glance at the history books shows that these apocalyptic prophecies did not come to pass. Improved access to education, birth control and abortion caused the global population growth rate to fall steadily over the next fifty years from above 2 per cent per year in 1970 to around 1 per cent per year today. The climax of the green revolution saw improved food production, particularly in developing nations. Even today, with its significantly increased population, the world is still able to produce enough food to amply feed everyone, although that food is not always distributed appropriately. We have reduced (but not completely eradicated) famine, whose typical underlying causes are now political instability, rather than global food shortages. In the 1960s, 500 of every million people in the world died as a result of food shortages. By 1990, this number had dropped to just 26 per million. The global death rate did not increase dramatically as Ehrlich suggested, but instead it decreased steadily from 13 per 1000 people in 1965 to below 8 per 1000 today.

Forty years later, when looking back at what he had written, Ehrlich admitted that many of the predictions he made in the book did not come true. In part, he attributed the incorrect projections to the fact that he had brought the problems to greater prominence through writing about them so dramatically, which, in turn, helped to inspire their solution. Despite recognising that his predictions did not come to pass, Ehrlich now believes he was 'way too optimistic', stating in one documentary interview, 'I do not think my language was too apocalyptic in *The Population Bomb*. My language would be even more apocalyptic today.' In effect, it seems that Ehrlich believes that he should have gone further and that his musings in *The Population Bomb* were, to some degree, *self-defeating prophecies*.

*

Self-defeating prophecies, as the name suggests, are the lesser-known antagonistic cousin of the self-fulfilling prophecies we met in the previous chapter: predictions which prevent themselves from coming true. Although less commonly employed than self-fulfilling prophecies, self-defeating prophecies are also a common literary trope. One of the most famous stories of a self-defeating prophecy was of that made by Jonah, one of the prophets of the Old Testament. The book of Jonah tells the story of how God had decided to destroy the city of Nineveh to punish its inhabitants for their pride, cruelty and dishonesty. He commanded Jonah to go to the city and prophesy its doom. Believing that his merciful God would allow the Ninevites to repent and thus avoid their destruction – making his prophecy self-defeating – Jonah decided that instead of embarrassing himself in Nineveh, he would flee. As we saw in Chapter 1, Jonah's flight eventually saw him spending three days and nights hiding out inside the belly of a big fish. After Jonah repents and agrees to go to Nineveh to prophesise the city's downfall, God gets the fish to regurgitate him and he promptly begins his mission. Upon hearing Jonah's fervent proclamations, the Ninevites repent. God spares them his immediate wrath and, as a result, brings to fruition Jonah's fear that his prophecy would be rendered false as a result of God's mercy. The risk of being thought a false prophet, when a valid but self-defeating prophecy is made, is often referred to as *the prophet's dilemma*.

The novella *The Minority Report* by Philip K. Dick and the film of the same name, starring Tom Cruise, are set in the near future in an alternate version of Washington, DC that has almost completely eliminated murder. Assisted by three 'precog' human oracles, able to predict homicides and led by Cruise's John Anderson, the elite

'Precrime' branch of the police is able to intervene and avert most murders before they happen. Each crime that a precog is able to predict is therefore a self-defeating prophecy. The fact that the pre-crime assailants never actually commit the crimes for which they are arrested leads Justice Department officials to question the ethics of the programme. Having witnessed a prediction in which he him-self is the perpetrator, much of the rest of the film revolves around Anderson attempting to avoid capture and, eventually, narrowly avoiding carrying out the murder that he was forewarned he would commit – another self-defeating prophecy.

In our modern world, self-defeating prophecies can have significant consequences. The early stages of the coronavirus pandemic provided one of the most divisive examples of a self-defeating prophecy. On 16 March 2020, researchers at Imperial College, led by Professor Neil Ferguson, released their now infamous *Report 9*. Among the charts and figures which illuminated the twenty-page document were two projections which stood out from the rest. The researchers suggested that if no action at all was taken to mitigate the spread of the virus, half a million UK citizens would lose their lives. Even under the government's prevailing *mitigation* strategy, the modellers suggested that 250,000 would die from the disease.[151] Faced with such stark projections, the UK government had no choice but to act. A week later, Prime Minister Boris Johnson addressed the British public on national television to announce a stay-at-home order, the closure of schools and a raft of other restrictive measures, which collectively would become known as *lockdown*.

Upon implementing the unprecedented social restrictions, daily case numbers began to fall almost immediately, as person-to-person contact was dramatically reduced. When cases were at their lowest in the UK, averaging around 600 a day in July 2020, the death toll stood at around 55,000 – around a tenth of what was predicted without any mitigations. Many mistook the low case rates, brought about by the ongoing restrictions and people's altered behaviour in the face of the deadly virus, for the end of the UK's epidemic. Consequently, many fringe groups and even mainstream media outlets began to ridicule the predictions of the Imperial College group. Ferguson was branded with the derisory moniker 'Professor Lockdown', in an attempt to lay the blame for what were characterised as unnecessary restrictions solely at the door of his group's seemingly incorrect predictions. In the autumn of 2020, *The Spectator* published an article which accused the Imperial College modellers of 'overestimating the number of people who were going to die', denouncing the projected figures as the number-one mistake in a list of 'The ten worst Covid data failures'.

The winter of 2020 saw a resurgence of Covid cases in the UK's still largely unvaccinated population. A further 95,000 deaths were added to what was already Europe's largest death toll from the first wave. Even in the autumn of 2021, while the UK death toll sat at over 150,000, the *Mail on Sunday*, in an article entitled 'Will the Covid doom-mongers NEVER admit they're wrong?' continued to suggest that Ferguson's half-a-million-deaths projection was not realised because he had overegged his forecasts. The *Mail* demonstrated a failure to appreciate the reductions in deaths that resulted from the restrictions and, latterly, the safe and effective vaccines. A few weeks later, the newspaper was forced to retract the article and apologise, admitting, 'We should have made clear that the 510,000 figure assumed no control measures'. It's unlikely that the *Mail* will be the last media outlet to mischaracterise a self-defeating prophecy as a failure of prediction, deliberately or otherwise.

It is not surprising that the mathematical modelling that is used to inform policy might make predictions that never come to pass, especially when the stakes are so high. It is precisely because they have informed policy that the terrifying predictions are averted. The fact that the assumptions on which the models were predicated have changed does not mean that those original projections were incorrect. Indeed, the UK's overall death toll of well over 195,000 by the summer of 2022, despite the restrictive mitigations that had been in place and the rollout of the vaccine programme, supports the magnitude of the original prediction. Ferguson himself suggested that, if anything, he believed his predictions were underestimates, having not taken into account the impact of an overwhelmed healthcare service which would have provided poorer treatment for patients had the virus been allowed to spread unmitigated.

There have been similar self-defeating impacts from predictions made across different branches of science and technology. Fear of the potential damage that might be caused by the millennium bug led to hundreds of billions of dollars being spent on updating computer systems to be able to deal with the potential calendar flaw. When 1 January 2000 arrived without significant disaster, many attributed the lack of incidents to the problem having been overhyped. The small number of issues that did occur, however, attested to the reality of the problem. The consequences may have been significantly worse had it not been for the widespread warnings of the dire consequences of allowing the bug to bite and the resultant self-defeating fixes that were applied.

When scientists began to understand the consequences of the developing hole in the ozone layer above Antarctica, they were quick to warn of the consequences for global agriculture and human health.[152] As a result of the severity of the potential repercussions if these predictions were to be realised, global leaders came together in 1987 to ratify the Montreal Protocol, which aimed to reduce the use of the damaging chlorofluorocarbons (CFCs) that were causing the hole. Very few people talk about a hole in the ozone layer anymore because the measures taken were enough to halt and reverse much of the damage that was done. The predictions that foresaw the potential problem were enough to startle the world into concerted action to prevent the disastrous predictions from ever coming true.

The Montreal Protocol was the first international treaty with the express aim of reducing global pollution. As I write, we stand on the brink of another global environmental disaster – the ever-increasing threat of global warming. We can only hope that current projections of the severe consequences of climate change are enough to stir our political leaders into meaningful action, reducing the amount of greenhouse gas in our atmosphere and turning the predictions of global temperature rises, along with the accompanying forecasts of the environmental and human costs, into self-defeating prophecies. Sadly, as we saw in Chapter 5, the solution to these multiplayer global 'games' is not always as simple as we might hope.

The underdog effect

Politicians themselves are not immune from the effects of self-defeating prophecies. Just hours before the 2016 US presidential election opened for voting, the Reuters/Ipsos 'States of the Nation' poll gave Hillary Clinton a 90 per cent chance of winning. Using aggregated polling data, statistician Nate Silver's 'FiveThirtyEight' website had correctly predicted the results of forty-nine of the fifty states in the 2008 election and all fifty in 2012. In 2016, FiveThirtyEight predicted Clinton would win 302 Electoral College votes to Trump's 235. Remarkably, despite having previously been so prescient, Silver's prediction (along with almost all the other official polls) was the wrong way around. Trump ended up with seventy-seven more Electoral College votes than Clinton.

Though many factors may have lain behind Trump's poor performance in the canvassing leading up to the election, it's possible that the lopsided poll results themselves played a significant role in the final outcome, effectively stymieing their own predictions. Seemingly assured of victory, many Clinton activists took their feet off the gas and some Democratic voters stayed at home. Conversely, the poor polling galvanised Trump supporters and encouraged marginals to vote with the Republicans.

Trump's surprise victory is an example of *the underdog effect*. Complacency in the favourite and determination in the longshot help to even out the odds and give the unfancied competitor a better chance than we might expect. In Aesop's fable, for example, the hare is so confident of his victory that he stops for a nap, while the unfavoured tortoise plods on slowly and steadily towards the finish line. The very fact that no one gives Sylvester Stallone's Rocky Balboa a chance in his world-title fight against undefeated champion, Apollo Creed, provides motivation for his underdog character. Combined with Creed's complacency – seemingly failing to do any preparation for the contest with a competitor he deems unthreatening – Rocky's determination to prove himself enables him to take Creed the distance, a feat no one has ever managed before.

Underdog stories are not restricted to the world of fiction. Many of the greatest underdog dramas come from the world of sport, and the root causes are exactly the same – overconfidence in the favourite combined with determination in the outsider. The game that ultimately determined the winner of the 1980 Winter Olympic ice-hockey competition saw the much-fancied Soviet team – winners of five of the last six gold medals – take on the young and inexperienced home team from the United States. While the Soviet team was composed largely of professional players steeped in international game time, the American side contained mostly amateurs, the most experienced of whom had only minor-league pedigree. The very fact that they had made it to the medal round was already a huge achievement for the US team. In contrast, the Soviets were expected, and indeed expecting, to win the tournament right from its outset. Just two weeks before it began, the Soviets had crushed the US team 10–3 in an exhibition match at Madison Square Garden. In hindsight, the Soviet head coach, Viktor Tikhonov, suggested that that victory 'turned out to be a very big problem', leading his team to underestimate the Americans.

On the day, the partisan crowd were four-square behind the home team. The US head coach, Herb Brooks, believed the Soviets were overconfident and capable of self-destruction. In motivating his team before the match, he suggested, 'The Russians are ready to cut their own throats. But we have to get to the point to be ready to pick up the knife and hand it to them.'

And so it transpired. Having taken the lead twice, the affront of being pegged back to 2–2 by the young upstarts led the Soviet coach, in a fit of pique, to substitute his talismanic goaltender, Vladislav Tretiak, widely regarded as the best stopper in the world. Whether intended or not, the message Tretiak's removal sent to the US team was that the Soviets felt they could brush aside the naïve Americans, even in the absence of their word-class, first-choice players. US team members would later cite the perceived slight as providing extra motivation to prove the Soviets wrong, whereas Tikhonov would describe the decision as the 'turning point' and 'the biggest mistake of my career'.

Despite falling behind again in the second period, the US team came back to level at 3–3 in the third and final period. With ten minutes remaining, the US team scored a fourth goal to give them a lead which they never relinquished. So implausible was the result, going down as one of the greatest upsets in sporting history, that it was christened the 'Miracle on Ice'.

However, we should be careful when telling underdog stories that we don't overgeneralise. Selection bias, driven by the fact that amazing, against-all-odds stories tend to stick in our minds, can make David beating Goliath seem to happen more often than perhaps it does in reality. On the majority of the occasions when a shoo-in contender meets a rank outsider, the favourite prevails – otherwise how would those two distinct reputations have been arrived at in the first place? Bookmakers do not make huge amounts of money by consistently mischaracterising winners as underdogs or vice versa. Nevertheless, there is verifiable experimental evidence that underpins the psychology behind the underdog effect.

Samir Nurmohamed researches the underdog effect at the Wharton School of the University of Pennsylvania. In real-world workplace experiments, he has found that people who perceived that they were not expected to be successful were more likely to do well on performance evaluations.[153] The research suggests that thinking of yourself as an underdog is strongly correlated with higher achievement at work – a form of self-defeating prophecy. In another lab-based study,[154] Nurmohamed's team gave 156 business students a negotiation task. Before their negotiations began, the students were told that the study's researchers had made one of three types of prediction about how they would perform in the task. Unbeknown to them, the supposedly bespoke predictions (high-expectation, neutral-expectation or low-expectation) were delivered to participants at random, independent of their previous track record on similar assignments. When the subsequent negotiation task was undertaken, there was a clear disparity in performance between the different prediction groups – those who had received the underdog-effect-inducing low-expectation predictions clearly out-negotiated the other two groups. Upon surveying the high-achieving underdogs, Nurmohamed and his team found that the desire to prove the predictors wrong explained much of the improved performance.

Interestingly, in separate follow-up experiments, Nurmohamed found that the credibility of those making the predictions or giving the feedback was crucially important for the success of the underdogs. If the subject believed that the person making the low-expectation prediction about them was unreliable, they were able to better employ the feedback as motivation to perform well. But participants who believed the source of the prediction about them was credible found it more difficult to overcome the seemingly authoritative low expectations.[155]

Negative feedback loops

Self-defeating prophecies and the underdog effect are examples of a more general phenomenon known as a *negative-feedback loop*. Negative feedback loops typically act to even the odds, to stabilise a system or to maintain the status quo. Self-defeating prophecies set in train a series of events which, eventually, render the events they predict implausible or impossible. For underdogs, the motivation derived from being an outside bet can combine with the complacency of the odds-on favourite to produce a result which is further from or even the opposite of what conventional wisdom might predict.

Remember that feedback loops are characterised by a signal, which triggers a response, or a series of responses, which ultimately end(s) up impacting on the original signal, closing the loop. In the previous chapter, we met *positive* feedback loops, in which the chain of responses served to reinforce the original signal, causing it to grow. In direct comparison, in a *negative* feedback loop, the responses triggered by the original signal may end up causing it to diminish. Changes to the status quo can be quickly undone by negative feedback.

Many of us will make use of negative feedback every day, perhaps without even knowing it. A ballcock in a toilet's cistern is part of a simple mechanical negative feedback loop used to regulate the tank's water level. When the toilet is flushed, the low water level is the input signal which causes the ballcock to drop. The lowered ballcock pulls open the inlet valve to which it is connected, allowing water to flow into the cistern. As the water rises towards the correct level, the ballcock floats upwards in the water, closing the valve and shutting off the flow, having restored the water to its baseline level. This is a brilliantly simple but effective use of a negative feedback loop.

Just as the word 'positive' in the positive feedback loops doesn't necessarily mean the quantity we are interested in is increasing, the 'negative' in negative feedback loops doesn't have to refer to a quantity that is decreasing either. Nor do the terms 'positive' or 'negative' infer value judgments on the impact of the consequences of the loops. Instead, the word 'negative' means opposing, describing the fact that the system undergoing the negative feedback loop is acting against the stimulus that first initiated it. While positive feedback loops tend to enhance changes to a system, negative feedback loops tend to dampen or buffer departures from the status quo. A more descriptive word for a negative feedback loop might be a *balancing* or *stabilising feedback loop*; when a stimulus unbalances a system in need of stabilisation, the chances are that a negative feedback loop will be called into action to redress the balance. Figure 8-4 illustrates two ways in which a negative feedback loop might typically act on a disturbance to a system, bringing it back to the status quo.

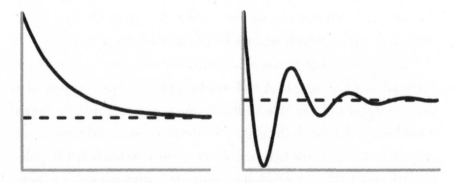

Figure 8-4: Negative feedback can bring a system quickly back to equilibrium (dashed line) after being perturbed (left panel). If there is a short delay between the signal and the response, negative feedback loops can give rise to oscillations (right panel).

Our bodies are masters of the negative feedback loop. For example, it's incredibly important for us to maintain a regular temperature. Too cold and we run the risk of hypothermia, whereas overheating can lead to heatstroke. Consequently, the part of our brain known as the hypothalamus monitors the temperature of the body extremely carefully and takes action to regulate it through the process of thermoregulation. In the cold, the hypothalamus can tell our hairs to stick out on end, trapping a layer of insulating air to ensure we lose less heat. It might also cause our muscles to contract and relax in rapid shivers in order to produce heat which raises the body temperature. Conversely, when we are too hot, we produce sweat, which allows body heat to dissipate efficiently by evaporating the moisture on the surface of the skin. Negative feedback loops are vital for homeostasis – the body's ability to maintain key physical and chemical attributes in the face of changing environments. They control everything from metabolism to fluid balance and from blood pressure to blood-sugar levels.

Anne Rice (whose novels you may remember clairvoyant Paula referencing in Chapter 1) was the best-selling author of *Interview with the Vampire*. At the age of fifty-seven she began to experience strange and unexplained symptoms. Having struggled with her weight for many years and having undergone several rounds of liposuction, she suddenly found she was dropping the pounds without really trying. While people around her complimented her new slimlined look, Rice herself was suffering with extended bouts of indigestion which seemed to accompany her condition. More concerning for the renowned author was the impact that the change seemed to be having on her ability to concentrate. She would sit in front of the blank page on her computer screen for hours at a time, unable to complete even the most basic descriptions of objects or characters. Rice sought the advice of a string of doctors who were able to rule out conditions like anaemia and cancer, but fell short when it came to putting their fingers on the root cause of her problems.

At an all-time low, Rice decided to tend to her 'spiritual health' by reaffirming wholeheartedly the Catholic faith she had neglected during the period of her rise to fame. Her physical health, however, continued to decline. Upon the renewal of her wedding vows with full Catholic pageantry, Rice described herself as being 'deliriously happy'. Over the weekend following the ceremony, however, Rice became genuinely delirious. She began to have trouble breathing and would only eat ice cream.

Although she remembers nothing of it, on the Monday morning following the ceremony, Rice called her assistants to her side and proceeded to tear her own clothes off. Her unusual behaviour and seeming incoherence led her assistants to call for medical attention. By the time a nurse arrived, Rice was unresponsive, having fallen into a coma. The paramedics who attended her took a blood-sugar (glucose) reading in an attempt to understand her condition.

To get the full picture of Anne's condition you need to know a little bit about blood sugar. Anything below 140 milligrams per decilitre two hours after eating is usually considered normal. A sustained level above 200 is considered indicative of diabetes. Levels above 600 can be indicative of the life-threatening HHNS (hyperglycaemic hyper-osmolar nonketotic syndrome).

Rice's reading was over 800. Doctors later told her that she was only fifteen minutes from cardiac arrest. She was diagnosed with diabetic ketoacidosis – the build-up of waste product ketones in the body, causing the alteration of the blood's pH – as a result of underlying and previously undiagnosed type-1 diabetes.

Sufferers of type-1 diabetes produce little or none of the hormone insulin, resulting in high blood-sugar levels. Sustained high blood sugar (known as hyperglycaemia) can cause life-threatening conditions like Rice's and damage vital body parts, including the eyes, nerves and kidneys. In contrast, sustained low blood sugar (hypoglycaemia) can cause clumsiness and confusion and, if not rectified, can lead to seizures or, in extreme cases, death. In non-diabetics, the pancreas and the liver work in tandem to maintain stable blood-glucose levels. When blood sugar is too high (after having a meal, for example), the pancreas secretes insulin into the bloodstream. Insulin transports glucose out of the blood and into the body's cells, where it can be used for energy. Insulin can also reduce glucose by converting it for storage as glycogen in the liver and muscles or by turning it into fats. After doing its job, insulin is broken down, so that it doesn't continue to remove glucose from the blood indefinitely. As blood-sugar levels fall, less and less insulin is produced, returning the body to appropriate and stable blood-glucose and insulin levels.

Alternatively, when blood sugar is too low, the pancreas secretes a different hormone known as glucagon. Glucagon has the opposite effect to insulin. It acts to break down stored glycogen into glucose and can also convert fats and amino acids to restore the blood's glucose levels. For most people, these two negative feedback loops acting antagonistically are enough to ensure that a regular and appropriate blood-glucose level is maintained. In undiagnosed diabetics, for whom natural insulin production is deficient, sustained high blood-sugar levels can be extremely dangerous, as Rice discovered.

Despite her brush with death, Rice learned to live with her diabetes for over twenty more years. Through careful eating and monitoring of the levels of glucose in her blood, she was able to recreate her own version of the negative feedback loop which regulates blood sugar. The injection of a carefully measured amount of synthetic insulin twice a day, when blood-sugar levels rise too high, is usually enough to maintain the status quo.

If there is a delay between the stimulus and the negative feedback which responds to it, we may find that our systems overshoot the desired equilibrium, eliciting a correction in the other direction. This, in turn, may overshoot, activating the original negative feedback and so on. Provided the feedback response is not overzealous and the delay between stimulus and response is not too long, then the system may oscillate its way towards the desired goal. The thermostats in our homes, for example, employ just such oscillating negative feedback to maintain a constant temperature in cold weather. If the temperature falls below the thermostat's setting, then the heating is activated until the room is brought back up to temperature. At the correct temperature, the thermostat may turn off the boiler, but the radiators, which are much hotter than the ambient temperature, may continue to pump their residual heat into the room, further raising the temperature past its desired setting. As the radiators, and hence the room, cools back down, the thermostat will register when it reaches the correct temperature, but the room may continue to cool before the boiler can heat the water required to fire up the radiators again. This series of diminishing-amplitude oscillations about the desired goal often characterises the behaviour of delayed negative feedback loops (as seen in the right hand panel of Figure 8-4).

In our house, I suffer these oscillating negative feedback loops almost every time I make pancakes (which is surprisingly often). In our kitchen, we have an old and fairly unresponsive electric hob, whose metal plates seem to take an age to heat up. Naturally, being impatient for pancakes, I turn the hob up to 'max' in order to get going with the frying as quickly as possible. The first pancake is invariably a write-off, as I try to flip it before it is properly cooked, the hob still warming up, despite what the dial indicates. The second is usually burned, as the hob heats up way past the right temperature before I realise it's time to turn it down. Alarmed at the smoke now issuing from the pan I turn the dial down way too low again. The first side of the third pancake usually cooks quickly as the hob cools down from a hot temperature, but the second side takes an age, as I've turned the temperature down too far. By the fourth or fifth pancake, I've usually alighted on a suitable cooking temperature in the pancake Goldilocks zone, but by then, the three hungry bears sitting at my table are getting restless.

Even if you're not the biggest pancake fan, it's likely that you've come face to face with these negative feedback oscillations when taking part in the mystical and involuntary dance known as the *sidewalk shuffle*: you're walking along the street and you see someone approaching you from the opposite direction, walking on the same side of the pavement as you. Being polite, you make a move to your left to avoid walking into them. At exactly the same moment, they politely sidestep to their right. You're back on a collision course. Your brain reacts to this new situation by telling you to move back to your original position. But there's a delay between your brain's instruction and your ability to act on it. While this is going on, your opposite number's brain is playing the same game, telling them to move back over. As you move back to the right, as if in a mirror image, they step back into your path. This continues until one of you consciously holds your nerve long enough to allow the other to react in isolation. Counterintuitively, you are sometimes forced to be selfish and hold your line in order to allow your dance partner to make the decisive move away in the nick of time. Alternatively, as often happens in reality, you are both forced to come to a complete and awkward stand-still so that one of you can pass the other before you both collide. The oscillations of the dance are the result of a delayed negative feedback loop, with each mirrored move to the right precipitating a correcting move to the left and vice versa.

If the negative feedback is strong enough, instead of damping down like the thermostat fluctuations or staying constant like the sidewalk shuffle, the amplitude of the response can grow out of control. On icy or waterlogged roads, this sort of growing negative feedback loop has been known to have fatal results. Oversteer occurs when a car's rear wheels lose friction, often while travelling round a bend. The back end tends to continue straight on, while the front turns, causing the car to point towards the inside of the bend. The temptation is to turn harder into the corner to counter the feeling that the back end is sliding off the outside of the bend, but the correct action is to steer the other way, into the direction of the slide, to allow the rear wheels to regain traction. In theory, this can allow the car to straighten up and the driver to regain control, but it's incredibly hard to get right. Often, the driver will panic and overcorrect, causing the back end to swing back out the other way. This might result in the car getting even more sideways than on the original slide but in the opposite direction. Overcorrecting for this second slide can again cause the back end to shift direction. The process, known as fishtailing (after the oscillating motion of a fish flapping its tail), can continue until the car spins completely, crashes into another vehicle or veers off the road. When the lag between the signal and the negative feedback response correcting it is large enough, it can effectively become a reinforcing loop which grows in time – each response being delayed just enough to coincide not with the signal it is attempting to cancel, but its opposite.

In the aviation industry, new pilots are taught to watch out for just such an *out-of-phase* negative feedback loop: the so-called *pilot-induced oscillation*. To hear former test pilot Tom Morgenfeld describe it[156]: 'It happens in any kind of system when the lags in the system get to be about the same as our own human response times'. And he should know; Morgenfeld was involved in a pilot-induced oscillation on a test flight of the prototype of Lockheed Martin's F-22 Raptor. Pilot-induced oscillations often occur when an aeroplane is coming in to land too hard or too steeply. The temptation for the pilot, as they see the ground rushing up to meet them, is to pull up hard to avoid a potential crash. As Morgenfeld put it[157]: 'The way you generally recover from those is to let go of the stick and everything will settle itself out, and you clean your trousers and come back and land. But when all you see is runway coming up at you, you're going to pull back.' By the time the pilot realises they've pulled up enough to avert the crash, they've already pulled up too much. Morgenfeld recalls:[158]

If I'd have pulled the nose up and taken my hand off the stick, the basic stability of the airplane would have taken over. But by virtue of the fact that I was trying to make a correction, there were huge lags in the system. I wouldn't see anything, and then I tried a little bit more. And then the first one would hit, so it would be an overcorrection.

When the pilot pulls up too much, the plane's angle of attack can become so steep that the aircraft dramatically loses speed and threatens to stall. If the pilot responds by trying to send the nose downwards again to gain speed, by the time they realise they've corrected enough they're already diving quickly towards the tarmac again, sometimes resulting in a faster dive than the one that precipitated the dramatic pull up in the first place. And so the cycle perpetuates and exaggerates itself until, as Morgenfeld recalls[159]: 'it got into a big oscillation . . . I bellied it in, and it hit real hard. That's what tore an engine loose. Then all that hot fuel, it severed the fuel line, so it started flaming.' After sliding for a mile along the tarmac with burning jet fuel spurting out of the rear, Morgenfeld's plane eventually came to a halt. Fortunately, he was able to quickly extricate himself, walking away relatively unharmed. The plane itself was a write-off. The negative feedback Lockheed Martin received in the media as a result of the crash was even more expensive than the negative-feedback-induced crash itself.

When negative feedback loops go awry, they can have the opposite effect to that which was originally intended. Sometimes the stimulus applied is intended to restore the status quo, but has the unexpected impact of sending the system spiralling off in the opposite direction. These uncontrolled increases or decreases, causing a reaction many times greater than the stimulus which triggered them, are precisely the boomerangs we met at the start of the chapter: the attempts to remove an image from the internet that inevitably leads to hundreds of thousands more hits; the super-injunctions that create such intrigue around the litigant that their identification ultimately becomes overwhelmingly in the public interest; the censorship attempts which turn bland books into best-sellers; the warnings which ultimately make it more desirable to take part in the proscribed activity; the incentives which precipitate a solution which makes the problem worse.

We must look out for these boomerangs in our everyday lives, being careful to think through the unintended consequences of our actions. Of course, this is easier said than done, otherwise they wouldn't be called unintended consequences. But, as we have seen, there are situations in which such backfirings are more likely to occur. If, like the Colombian military's reward system or US hospital league tables, targets are set for a proxy of the intended outcome, rather than the outcome itself, then we might expect people to try to reach those targets by any route – not necessarily the optimal route we might have hoped for. If we try to prohibit our kids from doing something we know is bad for them, we need to be careful that the ban itself does not inadvertently make the embargoed activity more alluring – fruit from the forbidden tree. Indeed, understanding psychological phenomena like reactance can give us the power to harness boomerangs for our own ends. Reverse psychology, when employed tactfully to induce the underdog effect, for example, can help us to get the best out of people. Here too, though, we must be careful that we do not go too far – that our use of the reverse-psychology boomerang does not itself come back to bite us by inducing the opposite of our intended effect.

In the last three chapters, we've discovered how nonlinear phenomena can upset our forecasts: from capricious reciprocals and seemingly deceptive square-cube laws to explosive positive feedback loops and counterintuitive negative ones. In the next chapter, we will combine much of what we have learned so far and consider what can happen when multiple different sources of nonlinearity combine to place fundamental limits on our ability to predict arbitrarily far into the future.

9

KNOWING YOUR LIMITS

So far, we have met a variety of nonlinear processes and discovered some of the difficulties we encounter when subconsciously assuming every relationship is linear. We saw specific examples of the real-world impact of nonlinear processes and the difficulties they can pose for prediction when we are not aware of them.

In this final chapter, we consider the reasons our judgment fails when it comes to thinking about nonlinear phenomena. In part, as we will see, we have hang-ups which extend from the linearity bias we met in Chapter 6. Our predisposition to see the world linearly puts us at risk of making the implicit assumption that everything will continue as normal, which leaves us particularly badly placed to cope with sudden change. On top of this, our verbal style of reasoning is just not cut out for extrapolating what will happen in complex and highly nonlinear scenarios.

In part, it is for this reason that I have been arguing that we need a more mathematical approach, in order to predict what will happen in these abstruse systems. Even this may not be enough, though. In the latter parts of this chapter, we will discover that *chaos* presents definitive horizons for even detailed mathematical models, beyond which we can no longer tell what the future will have in store.

Better than sex?

Acknowledging limits on our ability to predict arbitrarily far into the future is not to say we should give up on mathematical models altogether. Some accurate foresight, no matter how limited its scope, is usually better than none. Without mathematical models, we are left to the foibles of our own reasoning. Sadly, it is often the case that even when we think we are reasoning logically, our intuition lets us down.

And this is the problem. We all think we are so good at being logical that we rarely stop to step back and question our own reasoning. From a young age we learn to argue verbally – either out loud or in our own internal monologue. With verbal arguments we can convince ourselves that something must be the case, and sometimes we can convince other people, too, but there are many potential pitfalls.

A common aberration we are prone to make is known as the *single-effect trap*. We assume that every cause has at most one effect. If an action A decreases quantity B and a decrease in B also decreases C, then surely implementing A must lead to a fall in C. Makes sense. But what if the linear structure we have imposed by our verbal argument does not marry up with the full picture in the real world? What if we have failed to account for the fundamental nonlinear components of the system: the positive feedback loops (from Chapter 7) which might cause B to spiral out of control; or the self-regulating negative feedback loops (from Chapter 8) which keep C where it is; or the side effects that would make our strategy untenable. For example, what if action A also precipitates a rise in another quantity D, which then increases C? Executing A simultaneously leads to antagonistic effects which both boost and repress C. The only way to predict which effect will win out is to use a quantitative model. A simple, qualitative verbal model won't work.

This example feels a bit abstract and theoretical, but we don't have to look too far to find a real scenario in which the absence of quantitative foresight has led to unexpected problems. Abstinence-only sex education in the US received a significant increase in government funding under the Trump administration. The idea is that teaching children to wait until marriage before they have sex for the first time might lead to fewer teens having sex. If a fixed proportion of all teens stopped having sex, then there would presumably be fewer unintended pregnancies and sexually transmitted infection (STI) cases. Implementing A (abstinence-only sex education) reduces B (the number of young people having sex), and a fall in B leads to a fall in C (the number of unintended pregnancies and STIs), so according to our verbal linear model abstinence-only education should decrease teen pregnancy and STIs.

Abstinence-only sex-education seems to make sense, apart from one small catch. Although it might reduce the number of teenage sexual encounters, those teens who do still have sex and have not been taught about condoms and other forms of contraception are doing so far less safely than they would have been if they had received a comprehensive sex education. Abstinence-only sex education could thus lead to an increase in unprotected sex, which, in turn, would inevitably lead to a rise in teen pregnancy rates and STIs. A (abstinence-only sex education) increases D (the number of people having unprotected sex) which increases C (the number of unintended pregnancies and STIs). But which effect is more important?

In practice, the evidence for the significant drawbacks of abstinence-only sex education is fairly comprehensive. A 2015 study found that 'better-to-wait' sex education did not decrease the rates of STIs.[160] Nor does abstinence-only sex education lead to a decrease in teen pregnancy,[161] whereas comprehensive sex education *does* lead to a reduction in teen birth rates.[162] The peak in the abstinence-only movement in the US in 2006 coincided with the reversal in the longstanding decline of teenage pregnancy up to that point.

Our verbal model can only take us so far. If we want to know in advance how to avoid these sorts of backfirings, we need quantitative, dispassionate mathematical models to tell us whether we're going to throw a strike or a boomerang so that we can make informed decisions, in the full knowledge of how a situation will likely play out. Relying on verbal arguments to make policy, such as those advocated by the Christian right in the US, is likely to backfire.

Situation Normal: All Fucked Up

Verbal arguments, it transpires, are typically the manifestation of linear thinking: A leads to B leads to C. However, as we have seen in some of the preceding chapters, many of the most important and interesting phenomena in our world are not linear: from the homeostatic negative feedback loops which keep our bodies in check to the self-fulfilling placebo effects – the lies which make themselves come true. It is linear thinking which is responsible for *normalcy bias* – the tendency for people to believe that things will always continue to function in the future in the same way that they have in the past. When people think about what will happen in the future, they tend to imagine that things will either stay the same or will change linearly at the same rate they are currently changing.

Perhaps the best example of normalcy bias is our passage through life itself. Most of us start each day or each week believing it will be very similar to the one before. This is a reasonable assumption for most people most of the time, but for some, unfortunately, it will prove incorrect. No one anticipates that they will be involved in a traffic accident. No one young and healthy foresees a sudden cardiac arrest. We expect things to carry on as they always have done and we are always right, just up until the point that we are wrong.

This normalcy bias is evidenced by the number of people who haven't made a will. For some, the prospect of death seems so remote that planning for it by making a will seems absurd. Research from 2017 found that the majority (60 per cent) of UK adults hadn't made a will. Understandably, perhaps, the lowest rate was among eighteen to thirty-four-year-olds, just 16 per cent of whom had made one. But surprisingly, even among thirty-five to fifty-four-year-olds, arguably those most likely to have dependents and significant financial commitments, the rate was only 28 per cent. Even in the over-fifty-fives, who again, arguably, should be most aware of their time running short, over a third still hadn't made a will. The most common reason given for having so far failed to complete the task was a plan to do it later in life – a gamble on the assumption that things will continue as they are. I will admit that I am subject to this same complacency. Although I did make a will when my wife and I bought our first house – the first time I felt I had something substantial to leave to someone (not to diminish the significance of my collection of Oasis CDs) – I have not updated it since either of my children were born. I assuage my conscience by telling myself that I can always do it another day. I am banking implicitly on there being other days.

Normalcy bias is especially pernicious during natural disasters, causing people to ignore or minimise warnings of impending threats. Since at least the seventh century BCE, the people of Pompeii had been living in the shadow of Mount Vesuvius. Many inhabitants relied on the rich agricultural land surrounding the city, made fertile by the volcano's ash, for their livelihoods. Quality of life in Pompeii improved slowly but surely across the centuries. Under the Romans, Pompeii flourished, becoming a bustling city seen by many as a desirable place to live due to its proximity to the Bay of Naples and the rich farmland in the vicinity. By the first century CE, the population of the city had grown to between 12,000 and 20,000.

Much to the surprise of the city's residents, early one afternoon in the autumn of 79 CE, Mount Vesuvius erupted. Without warning, the volcano began to violently spew a column of volcanic debris, including ash, pumice and hot gases, up to 30 kilometres into the air. This phase of the eruption lasted several hours, allowing many inhabitants the opportunity to evacuate the city and head for safer climes. It is thought that upwards of 75 per cent of the population escaped during this early period. Around 2000 residents, however, did not leave. Some may have been too elderly or infirm to make the trip with the necessary haste. Some historians have suggested that the pumice raining down from the sky may have been enough

to put people off leaving their homes. Despite this potential deterrent, the evidence shows that the majority of residents did brave the
outside conditions in order to successfully escape the city. Other historians have suggested that reluctant citizens may simply have found
it difficult to believe what was happening, knowing that generation
after generation of Pompeiians had lived safely in the shadow of the
sleeping giant. After eighteen hours or so, the column of gas and rock
collapsed, ending the first phase of the eruption. The second phase
saw a flow of searing gas and rock descend the mountain at hundreds
of kilometres per hour. This first pyroclastic surge, followed soon after
by a second, buried Pompeii under a 6-foot layer of debris, asphyx-
iating and burning the residents who had chosen to remain behind.
They were victims of normalcy bias.

The title of this section 'Situation Normal: All Fucked Up',
often abbreviated to SNAFU, is usually used to characterise a sub-
optimal situation which is nevertheless the expected state of affairs.
Equally, though, it might be used to express the contrast between
the objective reality of a disaster like the eruption of Vesuvius and
the Pompeiian victims' subjective experience of it through the lens
of normalcy bias.

Perhaps the residents of Pompeii could have been forgiven for
their reticence to abandon the city they loved, having no former
experience of the potential impact of a volcanic eruption and no
overarching authority capable of quickly disseminating the urgent
message to evacuate. The same cannot be said of the victims of many
modern-day disasters. Even in the face of stark warnings given well
in advance of disaster scenarios – scenarios which have been widely
documented to have played out multiple times before – some people
still fail to react with the appropriate urgency. It has been suggested

that up to 70 per cent of people may display some degree of normalcy bias during a disaster.[163]

Denial is one of the key factors behind normalcy bias. People find the situation they are facing so out of keeping with what they are used to that they simply cannot believe it is happening to them. Contrary to what many of us might think, the fight-or-flight instinct does not necessarily kick in for everyone, even when they have accepted their situation. Some people find themselves in an incongruously calm state of *negative panic* or *behavioural inaction*,[164] while others frantically search for more information to confirm the warning they have received, squandering vital reaction time.

Meteorologists are perhaps foremost among the experts who find their warnings falling on seemingly deaf ears. Weather forecasters were aware, for example, of the formation of a tropical storm in the Atlantic Ocean by 20 October 2012. By 25 October, they were confident that it would make landfall at some point along the East Coast of the United States. Evacuation orders were issued across the Eastern Seaboard from the following day.

One of the most detrimental aspects of normalcy bias is that, even when we are given a reliable prediction of what the future has in store for us, one that hails from a trustworthy source, we may still fail to prepare appropriately for it. By the time Hurricane Sandy made landfall near Atlantic City, New Jersey, on the evening of 29 October 2012, nine days after it was first spotted, only 42.5 per cent of the state's residents in mandatory evacuation areas had actually left their homes.[165] Similarly, in New York City, towards which the storm was headed next, fewer than half of residents in the mandatory evacuation area 'Zone A' had left.[166] A total of 159 people across the East Coast of the US lost their lives as a result of Hurricane Sandy: 43 in New Jersey and 71 in New York state.[167] The leading cause of death was drowning.[168] Forty-five per cent of all the drowning deaths occurred in flooded homes in New York's 'Zone A', which was under a mandatory evacuation order.[169]

One survey of New Yorkers found that having witnessed traumatic events associated with 9/11 first-hand was associated with higher evacuation rates.[170] This finding suggests that having experienced the exceptional circumstances of a previous unexpected disaster does something to shake the 'it-couldn't-happen-to-me' mindset that is so prevalent in victims experiencing the denial stage of normalcy bias. Surprisingly, though, despite the devastation (Hurricane Sandy caused $65 billion worth of damage across the US, making it the second-most costly Atlantic hurricane in US history at the time[171]), one survey of residents of New Jersey found that only 54 per cent said they would actively prepare themselves in advance for another hypothetical storm in the future.[172] Well over half of those who claimed they would prepare, when asked what they would actually do, said they would 'gather information' or, worse still, suggested they would simply 'be prepared'.[173] The fact that the Eastern Coast had not experienced a hurricane as deadly as Sandy for forty years may have biased respondents to believe that, having recently lived through such an event, the need for preparation in the near future was minimal. If, as seems likely, global warming increases the frequency and severity of such storms, then these East Coast citizens may be in for another unpleasant surprise.

Weather, by accident or design

That we were able to foretell Hurricane Sandy's path and probable severity several days before it made landfall is an astonishing feat of prediction and one which would have been beyond our reach just a hundred years earlier. When Storm Eunice battered the UK in February 2022, bringing with it the strongest winds ever recorded in England, we knew about it at least four days in advance. In fact, the storm was predicted even before it had taken form out in the Atlantic Ocean. Unfortunately, three people died as a direct result of the storm, but that toll may well have been much higher without accurate forecasts.

While previously we relied on human experience to provide short-term forecasts of variable quality, modern science has rendered the accurate prediction of the weather several days into the future a minor everyday miracle. We have been trying for eons to come up with principles that allow us to know what the weather has in store for us. Some of these rules of thumb have a basis in science, while others are of dubious validity, only seeming to work when viewed through the lens of hindsight bias.

Take the old adage, 'Red sky at night, shepherds' delight; red sky in the morning, shepherds' warning.' In the New Testament's book of Matthew, a version of this lore is quoted by Jesus to doubters who come asking him for a sign from heaven: 'When evening comes, you say, "It will be fair weather, for the sky is red," and in the morning, "Today it will be stormy, for the sky is red and overcast". You know how to interpret the appearance of the sky, but you cannot interpret the signs of the times.' Several variations, the most notable, perhaps, replacing shepherds with sailors, appear in different areas of the globe, and the rule has its own version in languages other than English. In French, for example, '*Ciel rouge le soir, laisse bon espoir; ciel rouge le matin, pluie en chemin*' roughly translates as, 'Red sky at night, hope in sight; red sky in the morning, rain is coming'.

Its ubiquity and longevity suggest there is a kernel of truth behind the saying. And indeed, for many of the countries where the maxim persists, it is accompanied by some fairly solid science to explain it. The rule only generally holds when the predominant wind direction is west-to-east. This tends to occur at middle latitudes (between 23- and 66-degrees latitude both north and south of the equator), but not in the Tropics where the direction of the earth's rotation tends to mean the prevailing wind direction is often east-to-west.

To fully explain the phenomenon, as well as wind directions, we also need to know a little bit about the way sunlight interacts with our atmosphere. Although it appears white, visible sunlight is made up of different colours of light with different wavelengths, from the longest red wavelengths to the shortest blue and violet ones (you can see this when light is split up by raindrops in a rainbow). Usually, when sunlight interacts with the atmosphere, small air particles scatter the blue light (which is of a similar wavelength to the size of the particles) so that predominantly blue light enters our eyes, and we perceive the sky to be blue. However, when dust particles become trapped in the atmosphere in high-pressure regions, these larger particles tend to scatter the larger wavelengths of light – from the red end of the spectrum – towards our eyes, giving the sky a red or pink hue. In the evening, when the sun is setting in the west, this suggests a high-pressure system to the west, which will be blown towards us on the prevailing wind overnight, bringing fair weather for the next day. A red sky in the morning, as a result of the sun rising in the east, suggests that the high-pressure front has already passed over us from west to east, potentially making way for the wet weather that typically accompanies the arrival of a low-pressure system. The rule is not 100 per cent foolproof – the wind, for example, doesn't always come from the west, even in mid-latitudes – but it is supported by some sensible-sounding science.

Another well-known weather rule of thumb (in the UK, at least) is, 'Rain before seven, fine by eleven'. The rule may be less globally popular because its short rhyming structure only works in English, but more likely its lack of ubiquity stems from the fact that it is a less reliable and scientifically supported rule. In the UK, weather systems can move across the country relatively quickly, driven by the winds of the jet stream – sometimes, we genuinely do get precipitation starting before seven which has burned itself out by eleven. However, even in the UK, rain can – and often does – hang around longer than just a morning. As I write this on a miserable winter's day in Oxford, I can confirm it has been raining almost non-stop for two days now.

One of our favourite pieces of lore, that we like to look out for as a family if we're driving in the countryside, is cows lying down. The omen is thought to portend rain. Some proposed explanations suggest that cows can sense a change of air pressure or increased moisture in the air. They then lie down to keep a patch of grass dry for later grazing – or so the theory goes. One study went so far as to suggest that cows tend to stand up in warm weather, to expose a larger surface area, so as to cool themselves effectively.[174] The corollary to this suggests that, during the drop in temperature that often precedes a rainy spell, cows might lie down to preserve heat. These are plausible-sounding explanations for a phenomenon which, in fact, is not supported by science. Cows lying down is not a good predictor of imminent rainy weather. Any observed occurrences are likely to be pure coincidence combined with the confirmation bias that helps us to remember only the times when our proposed theory was correct and to forget about the times when the cows lay down and there was no rain. A Met Office survey revealed that 60 per cent of the UK public believe this piece of weather lore is a sure-fire predictor that rain is on the way.

In the German-speaking world, animal behaviour has also been harnessed in an attempt to predict the weather with, arguably, even less success. Eighteenth-century naturalists observed European tree frogs climbing up trees in sunny weather. Amateur meteorologists of the era overinterpreted the observation, supposing that this behaviour imbued the frogs with some sort of supernatural predictive power for fine weather. For a time, there was a fashion for keeping frogs indoors in jars fitted with miniature ladders, in case the weather-predicting desire to climb should arise. In reality, the outdoor frogs would climb trees in the wild when the sun was out in order to improve their chances of catching flies that tended to fly higher in warmer weather. Without the natural food source to encourage them, the jarred frogs had no motivation to climb, becoming a poor predictor of even the current weather, let alone future conditions. The term *wetterfrosch* meaning, literally, *weather frog* is still used in German-speaking countries as a derogatory term for weathermen. Female meteorologists, contrastingly, are sometimes disparagingly referred to as *weather fairies*.

Blame it on the weatherman

These deprecative nicknames are commonplace across the world. Over the years, weather forecasters have grown thick skins to deflect barbs like, 'Apparently being incorrect is a prerequisite for the job' or, 'Must be nice to still get paid when you're wrong more than you're right'. The irony is that this reputation for unreliability in forecasting the future not only undersells the extreme difficulty of predicting a complex system like the weather, but also underestimates the success rates of meteorologists in actually achieving this goal.

Of course, there have been genuine mistakes. Not even the most die-hard weather forecaster would deny it. On 4 September 1900, the Central Weather Bureau in Washington, DC spotted a 'tropical disturbance' moving northwards from Cuba. Believing it to be heading for Florida, they issued a storm warning for much of the state's west coast. By the following day, this had been upgraded to a hurricane warning for much of the Floridian and Georgian coasts, and storm warnings as far north as Kitty Hawk in North Carolina and as far west as New Orleans in Louisiana.

On the morning of 8 September, the residents of the city of Galveston situated on an island of the same name just off the coast of Texas, some 280 miles west of New Orleans and therefore not covered by the hurricane warning, awoke to swelling seas, but largely clear skies. High seas and minor flooding were nothing new on the low-lying island and, in the absence of any official warning, most of the city's 38,000 residents went about their business as usual. By the time the 'tropical disturbance' eventually made landfall later that day, it had become a category-4 hurricane, bringing with it a 15-foot storm surge. Given that the highest point in Galveston was less than 9 feet above sea level, the whole of the city was overwhelmed with flood-water. Many of Galveston's buildings were simply washed away along with the bridges connecting the island to the mainland. Wind damage in the city was also widespread. Winds of 100 mph were recorded before the Weather Bureau's anemometer (the device used for meas-uring wind) was blown away. Contemporaneous reports of flying debris, including bricks and timbers, suggest the wind speeds were much higher – consistent with the hurricane's category-4 ranking. The next day, when the worst of the storm had passed, the surviving residents were left counting the costs. Ten thousand were left home-less and around 8000 citizens had been killed. Among the dead were Cora Cline – the wife of Galveston's chief meteorologist, Isaac Cline – and their unborn child. Cline himself almost perished in the flood, paying a heavy price for his failure to raise the alarm about the storm until it was too late. The Galveston hurricane remains the deadliest recorded natural disaster ever to befall the United States.

These stories of unmitigated disaster resulting from a failure of prediction are thankfully rare, especially in the modern era of fore-casting. But one missed hurricane stays in the memory far longer than the correct foretelling of any number of sunny days. Notably, incorrect predictions also stand out more than the correct predictions of potential disasters which subsequently enable those disasters to be averted. The accurate prediction of the path of 2011's Hurricane Irene, combined with the effective preparation afforded by the early warning, meant that the storm resulted in far fewer deaths than it might otherwise have done. Despite being more recent than 2005's Hurricane Katrina, the lower toll of death and destruction means that fewer people remember Irene than her devastating older cousin.

Many of those who do remember Irene recall the warnings around her approach as overblown which, as we saw in Chapter 8, is the classic, thankless repercussion of a successful but self-defeating prophecy.

Our tendency to overestimate the probability of events which are more readily recalled and to underestimate that of more mundane events can be categorised as a type of availability bias. We first met availability biases in Chapter 1, when we discovered the recency effect (responsible for the Baader-Meinhof phenomenon) that keeps freshly acquired information easily accessible at the forefront of our minds. Another phenomenon which mediates availability, the ease with which we remember one event compared with another, is known as *salience bias* – our tendency to ignore less remarkable events in favour of those that are more prominent and emotionally impactful.

Availability bias explains patterns in the way people purchase insurance. Take disaster insurance, for example. We tend to purchase insurance based not on the true risk of a disaster, but on our perceived risk. People tend not to renew their policies if there hasn't been a disaster for a while to remind them why they need them – their perception of the risk goes down. The magnitude-6.9 Loma Prieta earthquake which struck central California in 1989 caused almost a billion dollars' worth of insured damage.[175] The stock prices of insurance companies might have been expected to have taken a big hit based on the huge sums of money they were forced to pay out. Instead, insurance stock prices increased in the immediate aftermath of the earthquake.[176] Investors realised that the recent earthquake would cause an availability bias. People's heightened perception of risk[177] would lead to an increase in sales of earthquake-insurance policies, which would be more than enough to cover the payouts.[178]

A similar situation saw the number of flood-insurance policies taken out under the US National Flood Insurance Programme increase slowly by between 0 and 4 per cent each year between 2001 and 2009, with the exception of 2006. In 2006, the number of active flood-insurance policies increased by over 14 per cent.[179] What happened in 2005 to precipitate this sudden increase? Hurricane Katrina, of course. With such prominent coverage of flooding appearing on the news for so long, people's perception of the risk of flooding increased, despite the fact that they were at no greater risk than before. The frequency of hurricanes and the distribution of their locations had not changed. Hurricane Katrina did not make the risk of flooding any greater or any less, it just made an example of flooding more available.

Salience bias certainly plays a role in explaining the perception that weather forecasts are wrong more often than they are right. We remember mistakes like the Galveston disaster, whose outcomes were terrible, much more than we remember Hurricane Irene whose loss of life was largely mitigated by accurate forecasting. The salience bias associated with an incorrectly forecasted hurricane in the past can override the recency effect associated with a contemporary but well-predicted counterpart.

Another piece in the jigsaw puzzle of our weather misconceptions is our fundamental misunderstanding of the uncertainties associated with the forecasts. We would prefer it if our weather forecasts told us with 100 per cent certainty that it will rain tomorrow. That way we don't get lumbered with our umbrellas unnecessarily. But weather forecasts don't work like that; they are inherently uncertain. In the last decade or so, meteorological organisations have bitten the bullet and decided to try to communicate that uncertainty to the public. You will probably be familiar with forecasts that predict, for example, a 40 per cent chance of rain for your local area. These estimates are often known as *probabilities of precipitation* or *PoPs* for short. Unfortunately, there is some ambiguity about what these PoPs actually mean. Does it mean that 40 per cent of the forecast area will experience rain the following day? Perhaps it means that it will rain for 40 per cent of the day across the area. Perhaps it means that all of that locale will experience rain with a 40 per cent probability – that is to say, looking back over the history books, 40 per cent of the days with conditions like tomorrow saw rain, but 60 per cent did not.

This last option is probably closest to the correct interpretation, but the true meaning is slightly more complicated still. For many weather forecasts, the PoP for the next day should be interpreted as the probability of rain tomorrow in a given area, multiplied by the proportion of that area that will see rain. So, for example, if a weather forecaster is certain it will rain in my native city of Manchester tomorrow but only across three-quarters of the city's area, then the PoP they present should be 75 per cent. Similarly, a 75 per cent PoP might be arrived at from a forecast which suggests that there is only a 75 per cent probability of rain, but if it does rain, the whole city will get wet. You can see how people might misinterpret these uncertainties. In the first scenario, if you live in the quarter of Manchester that doesn't get rained on, then you might say the forecast was incorrect. In the second scenario, if the rain doesn't come (as it was predicted not to on one quarter of such occasions), Mancunians might again characterise it as incorrect, despite the forecast being valid in both cases. Anecdotally, I never found these refined forecasts so useful when I lived in Manchester anyway. Everyone resident in the city knows it has just three types of weather: either it's raining, it's about to rain or it's just rained.

Frustratingly for meteorologists, even attempting to give a nuanced view by communicating the uncertainty of the chances of rain might be viewed by the general public as a cop-out that covers all the bases. When the UK's Met Office first started giving numerical probabilities of precipitation in their forecasts, the *Daily Mail* suggested, 'Anyone who's ever been caught in an unexpected downpour may feel they [weather forecasters] have simply come up with a way to deflect blame when they get it wrong'.

Part of the problem is that we are encouraged to think in binary ways about the weather – either it will rain or it won't. As well as giving a numerical probability of rain, many weather apps also typically display an icon to illustrate the expected weather visually. Naturally, this requires forecasters to set a threshold on, for example, the probability of rain, above which the 'cloud with the raindrop' symbol will be displayed in preference to the 'grey cloud' icon. Even when ignoring the symbols and looking at the numbers, it's too tempting optimistically to read a 20 per cent chance of rain in your specific area as low and to assume it won't happen. On the one-in-five days when there is a downpour and we get caught in a shower without our coats, we might be angry that the forecast didn't predict rain with more certainty.

Interestingly, some commercial weather forecasters in the United States are only too aware of our rounding biases and our preferences for being over- rather than under-prepared when it comes to rain. By comparing precipitation predictions to the frequency with which rainfall actually transpired, researchers found that when the true chance of rain was low, the predictions of commercial forecasters were biased towards an increased chance of rain.[180] A potential reason behind this so-called *wet bias* is precisely because forecasters understand our tendency to round down low percentages. They appreciate that when many of us see a chance of rain that is 10 per cent or smaller, we implicitly assume that it will not rain. By deliberately upping the probabilities to 20 or 30 per cent, forecasters make us take the chances of rain more seriously, reasoning that we won't be disappointed if they predict rain and it turns out not to happen. The same research also found that forecasts were skewed artificially away from PoP predictions of 50 per cent, presumably to avoid the ambiguity associated with such forecasts.[181] Needless to say, these biased predictions decrease the accuracy of forecasts in general and do little to improve the public's confidence in them.

The same thing happens on a larger, but arguably more consequential and more legitimate scale with the prediction of tornadoes, hurricanes and blizzards – natural weather disasters. Forecasting these phenomena is a high-stakes game of chance. Authorities in charge of disaster and emergency planning often work from weather forecasters' reasonable worst-case scenarios – disastrous projections, which may be unlikely to unfold, but not so unlikely as to allow them to be dismissed out of hand as impossible – because they simply cannot afford to miss such an event. It is typically considered better to overreact, causing unnecessary disruption and inconvenience, than to underreact and to fail to avert preventable loss of life.

In March 2017, when the National Weather Service predicted blizzard conditions for residents of the East Coast of America, including New York City, 20 million people prepared for the worst. Headline news stories heralded predictions of wind speeds of up to 60 miles per hour and between 1 and 2 feet of snow in the Big Apple – rivalling the Great Blizzard of 1888. Over 7000 flights were cancelled, affecting 350,000 passengers. Hundreds of train journeys were cancelled or suspended. Businesses and attractions closed down in preparation. New Yorkers were encouraged to avoid all unnecessary travel. President Trump even cancelled a meeting with the German chancellor, Angela Merkel. The precautions taken as a result of the forecasts disrupted everyday life and led to short-term but substantial economic damage.

Ultimately, when the storm passed over New York, fewer than 7 inches of snow fell. Many of the more stringent preparations had been unnecessary. The National Weather Service, responsible for the incorrect predictions, took a beating on social media, with many users describing the forecasts as 'a bust'. Chris Christie, the Governor of New Jersey, was so angry about the incorrect forecast that he ranted, 'I don't know how well we should be paying these weather guys. I've had my fill after seven and a half years of the National Weather Service to tell you the truth.'

In fact, the high likelihood of reduced snowfall totals in New York was known by the National Weather Service the evening before the storm was due to drop. Out of an abundance of caution, the decision was taken not to cut the snow forecast. Greg Carbin, the Weather Prediction Center's Chief of Forecast Operations, explained that a last-minute downgrade might have given people across the affected states the false impression that the storm was no longer a threat. Known as the *windscreen-wiper effect*, dramatic flip-flopping between vastly different forecasts is widely believed to undermine public confidence even more than incorrect predictions.

The difficulty in forecasting the snow depth for New York was because the city was predicted to lie close to the storm's rain–snow line – the front separating cold Arctic air and warm, moist Atlantic air. Ultimately, the predicted amount of precipitation did fall on New

York, but much of it as rain and sleet, rather than snow, as the rain–snow line shifted from its predicted course. Inland areas of New York state did, indeed, receive the predicted 1–2 feet of snow, with some areas receiving over 4 feet.

It is reasonable, and even sensible, for forecasters of almost any phenomenon to present a range of possible scenarios with accompanying uncertainty, rather than to present a single scenario as if its occurrence is a complete certainty. Much of the discord between what we perceive was predicted and what actually occurs arises when the communication of the uncertainty around the forecasts is lacking, or worst-case scenarios are interpreted as being more probable than they truly are – sometimes as 100 per cent certainties. We long so much for this unambiguous assurance that we frequently forget or ignore the nuance that usually lies behind these predictions.

Some proportion of the weather forecasters' disrepute may be self-inflicted, then – borne out of an abundance of caution, preferring to reduce the risk of missing a potential natural disaster at the cost of increasing the rate of false alarms. To some extent, their lousy reputation may derive from our fundamental inability to appropriately parse probability (a problem we encountered first in Chapter 2), but some of the meteorologists' ignominy is a result of the general under-appreciation of how fundamentally difficult a problem predicting the weather is.

The demon's in the detail

One hundred and fifty years ago, the science of weather forecasting was in its infancy. In the early days, the method of prediction involved making some atmospheric observations and poring over the historical records to find days with similar characteristics. The weather that was recorded to have followed the closest-matching day in the record books was used as the forecast. Unsurprisingly, this crude, *ad-hoc* approach to predicting an extremely complex system like the weather didn't yield much success.

Despite the equations which govern atmospheric physics being well known, for much of the twentieth century there was no way to practically solve them with the accuracy and detail needed to make reliable and useful forecasts. This all changed in the 1950s with the advent of supercomputing. In the computer representations of the atmosphere, the surface of the globe could be partitioned into a grid on which the variables that determine the weather (wind speed, air pressure, humidity, temperature, etc.) could be modelled. The governing equations provided a formula which allowed the machines to calculate the weather in one grid point at a particular time from the conditions in the neighbouring grid points at earlier times. Given an initial snapshot of the atmosphere at the current time, the supercomputers could iterate the models fast enough to spit out forecasts that were of practical use to inform decisions about how best to react to those forecasted future conditions. In contrast, had human forecasters sat down to crunch the numbers by hand, the forecast would have been well out of date before it was even finished.

As computing power improved, so did the forecasts. More atmospheric variables could be included, and the grids could be made finer to give higher-resolution predictions. Solving the equations became fast enough that, rather than just running the models once, the supercomputers could run them multiple times. Adding small differences to the initial atmospheric snapshot in each repeat, to mimic the inherent uncertainty in those measurements, allowed forecasters to model how these differences propagated, altering the forecasts over time, enabling the quantification of the uncertainty around the predictions. The more times the simulations can be run, the more confident we can be about the forecasts.

Over the years, the accuracy of predicted weather patterns has improved with the advancement in supercomputing power and simulation technique. An approximate rule of thumb suggests that every decade, the distance into the future for which we can forecast with a given degree of accuracy improves by a day.[182] That is to say that a four-day forecast today is as accurate as a three-day forecast was ten years ago. Where we peered four days into the future thirty years ago, we can now look a week ahead with the same degree of confidence.

Despite these improvements, there will always be fundamental and inherent uncertainty associated with predicting the behaviour of a complex, nonlinear system like the weather. Part of that uncertainty stems from the approximations we have to make when building the requisite models. We don't represent all 200,000,000,000,000,000,00 0,000,000,000,000,000,000,000,000,000 (two hundred tredecillion) molecules in the atmosphere using molecular-scale models. It's just not possible with current computing facilities, nor is it likely to be possible any time soon. At the moment, models at this 'molecular' scale are capable of simulating a few million atoms[183] for a few hundred microseconds[184] at a spatial scale of a few hundred millionths of a metre. In contrast, weather forecasts require models to be able to handle global spatial scales of tens of thousands of kilometres and time scales that stretch to at least a week. Models which represent each molecule individually are simply not feasible at this scale. Instead, the variables that are fed into the computer need to be coarse-grained representations of fine-grained reality to make the models simple enough to run efficiently. Computationally cheap though these coarse models may be, their cost lies in the fact that they can only ever be cartoon representations of reality – passable likenesses perhaps, but never photo-level realism.

The theoretical idea of modelling reality at the finest imaginable level of detail is not a new one. In 1814, the French mathematician Pierre-Simon Laplace (an early proponent of Bayes' theorem, who we first met in Chapter 4), dreamed up a thought experiment that subsequently became known as *Laplace's demon*.[185] Laplace proposed a super-intellect – the demon – that at a given moment in time would instantaneously know everything about the universe – the position and momentum of every particle and the rules according to which they interact with each other. He reasoned that, using the laws of classical mechanics, the positions and momenta of these interacting particles could be evolved forwards in time, making the whole uni-verse entirely predictable from that point onwards. His proposition, very much like the problem of predicting the fine-grained behaviour of the atmosphere in excruciating detail, is, of course, unrealisable in practice. What computer could encode all the information required, let alone complete the computations?

As a thought experiment, though, Laplace's suggestion has pro-found consequences for how we think about free will, and in quite a different way to the way that I felt my freedoms were limited by the Aaronson Oracle in Chapter 3. Then, if you recall, a computer was able to predict which of two keys I would press next with roughly 60 per cent probability. Better than the 50 per cent that would be predicted purely by chance, but hardly the level of forecasting accu-racy that would convince us to bet our houses on the outcome of the prediction. Laplace's thought experiment, in contrast, suggests that all our current and future actions are already determined – that we are, in fact, just automata, acting out the playbook set in motion at the start of the universe. If the demon were real, it would be able to predict which of the two keys we would press with 100 per cent certainty. If the central tenets of Laplace's theory hold true, then free will is just an illusion – a result of our inability to specify our starting conditions accurately and thence to compute forwards in time.

Perhaps, though, it doesn't matter whether we genuinely have free will or just the illusion, since we also lack the ability to tell the difference. Not only are our computational resources insufficient, but there are fundamental limitations to how accurately we can measure the positions and impulses of particles: not least those posed by the fundamental indeterminacy encapsulated in *Heisenberg's uncertainty principle*. Most pertinent at very small spatial scales, the uncertainty principle of quantum mechanics suggests we can never pin down both the position and momentum of a particle at the same time with absolute certainty: if you know exactly how fast something is going, you can't know exactly where it is.

The inability to accurately determine the initial snapshot of the atmosphere, both theoretically because of Heisenberg's uncertainty principle, and practically because of the sheer complexity and size of the system, is an important part of the second main reason why weather forecasts become inaccurate – chaos.

Total chaos

In the mathematical sense (as opposed to the commonly understood usage to mean disarray and disorganisation), chaos is often characterised[186] by what mathematicians refer to as *sensitive dependence on initial conditions*. This means that the behaviour of two otherwise identical chaotic systems, initiated with an extremely similar (but not exactly identical) initial set-up, will eventually diverge from each other if we watch their evolution for long enough.

You might think that in order for a system to exhibit chaotic behaviour it needs to be extremely complex. But this isn't the case. A frictionless double pendulum is a classic example of a simple chaotic system. Imagine the single pendulum of a grandfather clock – a solid metal rod with a bob attached at the bottom. The behaviour of the single pendulum is understood extremely well. It doesn't exhibit chaos and behaves very predictably – so predictably, in fact, that we can (and do) almost literally set our watches by it. Pin another pendulum to the bottom of the first, however, and the whole situation changes. All of a sudden, despite the fact that the system is still simple enough that we can completely characterise its behaviour mathematically, the double pendulum can now exhibit chaos.

The upper panels of Figure 9-1 each show the path of the end of the pendulum in two scenarios for which it has been released from a near vertical upside-down position (shown in grey). The initial angle between the pendulums in the two panels differs by less than one tenth of a degree. After initially behaving quite similarly, the trajectories of the pendulums diverge in under five seconds. As you can see in the bottom two panels of Figure 9-1, after twenty-five seconds, the patterns generated by following the paths mapped out by the end of the pendulum are radically different in the two scenarios, despite each having an extremely similar initial condition.

Figure 9-1: The top two panels show the first five seconds of the trajectories (black) of the end of two almost identically initialised (initial positions given in grey) frictionless double pendulums. The trajectories appear similar initially but have clearly begun to diverge even after five seconds. By twenty-five seconds (lower two panels), the trajectories mapped out are completely different. This time the final position of each double pendulum is shown in grey.

This sensitive dependence on initial conditions has significant ramifications for our understanding of chaotic systems. When building models for the prediction of systems like the weather, chaos means that despite the fact that the laws governing a given system might be well understood, if there are even small uncertainties in the starting positions that we feed into our model of the system, after some time, the evolution of the real system will diverge significantly from that predicted by the model.

You can hear chaos for yourself in your own kitchen.[187] Place an upturned baking tray in the sink and turn the tap on ever so slightly. When you've opened it enough, you should find that the drops tap out a regular metronomic rhythm on the tray – the sound of a leaky tap that can drive us to distraction. Turn it on a little bit more, however, and you will find the drips begin to fall with an irregular pattern, not forming individual drops until well away from the faucet opening. If you close your eyes and listen carefully, the unpredictability of the sound is almost mesmeric.

Chaos appears frequently in the world around us, often without our realising it. It is thought to characterise the splish-splash of water falling from a fountain and the variations in animal population sizes.[188] If you've ever played pool or watched it on TV, you'll appreciate that no matter how fine the player taking the first shot is, they are never able to precisely replicate a break off. But as a pool-playing friend described it to me, 'This isn't a bug, it's a feature – it keeps the game interesting'. Minor alternations in the speed and the angle of the shot, and in the way the triangle of balls is racked up, mean that every break is different. Small differences evolving into larger ones in a way which makes the outcome practically unpredictable is one of the essential features of chaos.

Although the locations of the pool balls can't be determined ahead of time, this is not the same thing as them being random. We have equations which describe the evolution of each of the chaotic phenomena in the above paragraph – the growth of an animal population, the dynamics of fluid flow, the mechanics of interacting pool balls. The macroscopic laws of physics allow no room for the spontaneous evolution of randomness, specifying exactly how the pool balls should move. If we knew the initial conditions of the system exactly, we could predict the future evolution with complete certainty. But small errors in the initial conditions mean that the predicted future trajectory in our model will diverge from the true one after a short period of time in a chaotic system. This is where the randomness comes in. The uncertainty in our measurements of the initial condition is propagated and amplified.

As we saw in Figure 9-1 with the evolution of the double pendulum, two systems starting with near-identical initial conditions may stay on almost the same course as each other for a short time, but if allowed to evolve for long enough, they will eventually diverge, going on to map out radically different paths to each other. Practically, this means that many complex systems, like the weather, cannot be predicted with much accuracy beyond a given time horizon. For the weather, with our current accuracy of measurement and modelling, this horizon is between one and two weeks.[189] After that time, forecasts are typically outperformed by looking at the historical weather records and averaging the probabilities of a given weather phenomenon on that same day in previous years.

Although the first daily weather forecast was not published until 1861 in *The Times*, the history of weather forecasting goes back much further. The various pieces of 'lore' and rules of thumb we have met already are a strong indication of our long-held desire to predict what the weather has in store for us. Civilisations as far back as the Babylonians had noted what they saw as meaningful connections between weather phenomena and the positioning of celestial bodies. The ancient Greeks were among the earliest civilisations evidenced to have actively used astrology to attempt to forecast the weather. Medieval astronomers and astrologers drew on ancient Greek, Indian, Persian and Roman knowledge to replace more primitive (but arguably no more inaccurate) forms of forecasting with a new, formalised field of scientific exploration known as *astrometeorology*. Now considered a pseudo-science, its central tenet was that the celestial bodies could influence and predict the earth's weather. Astrometeorology held sway for a surprisingly long time, even after some other forms of astrology (for example natal astrology – the branch we probably think of first when someone mentions astrology – which suggests that the position of celestial bodies at the time of your birth can influence your path through life) had fallen out of scientific favour.

That it endured for so long is perhaps unsurprising given what was known at the time about the influence of celestial bodies on earthly phenomena. The sun's heat and light had long been known to have a significant impact on the earth's climate. The moon, too, was understood to influence a phenomenon as momentous as the shifting of the tides. Why then, it was reasoned, would other more distant bodies not also dictate natural phenomena on earth? Even until the late sixteenth and early seventeenth centuries, among astrometeorology's adherents were numbered some of most renowned astronomers of the day, including Tycho Brahe and Johannes Kepler. Tycho, in particular, was influential in establishing the widespread practice of keeping weather diaries, which he hoped would supply vital data to improve future astrometeorological predictions. Ironically, such detailed weather records would help provide the data for the establishment of what we might think of as modern scientific meteorology in the seventeenth century, which led, subsequently, to astrometeor-ology's fall from acceptance in scientific circles.

By the late seventeenth century, the foremost astronomers of the day had turned their attention away from forecasting the weather towards the problem of predicting the positions of the celestial bodies themselves. In November 1680, Sir Isaac Newton observed a comet he had been tracking disappear behind the sun. A few weeks later, in December of the same year, another comet reappeared on the other side of the sun. Newton hypothesised that this must be the same celestial object, but that its path must have curved substantially for it to have passed so quickly behind the sun. For the path to bend so sharply, Newton realised that there must be an unseen force acting on the comet – a force he called gravity. Despite this revelation, Newton couldn't quite reconcile the motions of the comet he had observed with his new laws of motion and gravity.

His friend Sir Edmund Halley, however, picked up the trail Newton had let grow cold. Accounting for the gravitational pull of Jupiter and Saturn, using Newton's own laws of motion and universal gravitation, Halley was able to predict that his now epon-ymous comet would reappear in 1758. Although neither man lived to see it, the comet dutifully came into view on Christmas Day of 1758 to great fanfare. This first demonstration that celestial bodies other than the planets orbited the sun was a clear validation of the predictive powers of Newton's laws. The motions of the objects in the solar system were, it seemed, as regular as clockwork – the very definition of predictability.

By the mid-1800s, scientists armed with Newton's laws were growing overconfident. The successes of Newtonian mechanics in predicting the existence of Neptune in 1846 – not by observing it directly, but by using mathematics to determine it must be there[190] – was yet another feather in the cap for the predictive powers of science. Surely, if mapping the future movements of the heavenly bodies themselves was now within the mathematicians' realm of capabilities, then no physical problem could prove too great. Mathematicians and physicists began to believe in Laplace's forecasting utopia – that, given the precise initial conditions, they could project arbitrarily far into the future in almost any scenario.

Not everyone accepted the dogma so readily, though. In 1885, in honour of Sweden's King Oscar II, Swedish mathematician Gösta Mittag-Leffler and Russian counterpart Sofia Kovalevskaya set a challenge to the world's scientists. Anyone who could prove what became known as the n-*body problem* – that *n* objects attracting each other according to Newton's universal law of gravitation would remain stable over time – would be awarded a gold medal and 2500 Swedish kronor. Despite the abstract formulation of the problem, what Mittag-Leffler and Kovalevskaya were really asking about in a practical sense was the stability of the solar system. Would the eight principal planets continue to orbit regularly around the sun for ever or could one of them fly off unpredictably out of the solar system?

After three years of waiting, a manuscript that purported to solve the problem finally arrived. Three hundred pages containing the foundations of entire fields of new mathematics were submitted by celebrated French mathematician Henri Poincaré. In his dossier, he simplified the *n*-body problem to a three-body problem in which two large masses orbiting each other (like a binary star system) were joined by a third smaller mass interacting with the other two. When he studied the evolution of these three bodies, he found, indeed, that the system would remain stable. Although he hadn't completely solved the original problem as posed, for his work he was nevertheless publicly awarded the gold medal and the 2500-kronor prize.

But just as the pioneering work was about to be published in Mittag-Leffler's own journal, *Acta Mathematica*, he received a telegram from Poincaré telling him to stop the presses. Poincaré had discovered a mistake in his work – a mistake so fundamental that it completely changed his conclusions. Now, instead of stability and predictability, Poincaré was suggesting that one of the bodies could readily be jettisoned far from the system. He found that even small alterations in the initial positions or the masses of the three interacting bodies could dramatically change the results of his calculations. Small rounding errors could quickly expand in size. Turning the received wisdom on its head, his revised work concluded that the solar system exhibited *dynamical instability* and was therefore too complicated to ever succumb to accurate forecasting arbitrarily far into the future. Poincaré's revised findings effectively foreshadowed the discovery of chaos decades later. It is this version of his work which was eventually published in December 1890, over five years after the inception of the prize, and which is still accepted as correct to this day.

Although the laws Newton devised (and, more recently, the theory of general relativity for more accurate calculations) can indeed be used to give seemingly accurate predictions of the future configuration of our solar system, the motions of these celestial bodies are, as Poincaré discovered, actually chaotic. This planetary chaos, however, only becomes apparent on relatively long timescales – a chaos-horizon of the order of tens to hundreds of millions of years. Knowing the current positions of the planets to a good degree of accuracy will allow us to predict their positions well for a few million years, but eventually, after a long time, a planet might be found on completely the opposite side of the solar system to where today's calculations would locate it. This isn't because of any randomness in the dynamics of the planets – their motion is well understood and described by Newton's laws of motion and universal gravitation – but is a result of the fact that planetary motion is a chaotic system. Chaos is a fundamental feature which limits the predictability of many complicated non-random phenomena. As mathematician and meteorologist Edward Lorenz supposedly characterised it, 'Chaos: when the present determines the future, but the approximate present does not approximately determine the future'.

Indeed, it was Lorenz's attempts to forecast the weather in the 1960s which led directly to his discovery and characterisation of what we now consider to be the mathematical subject of chaos theory.[191] He had built a relatively simple (by today's standards) model of the earth's atmosphere which he was attempting to solve using a fairly primitive (again by today's standards) computer. The computer would print out the results of the model at regular time points as a series of numbers describing the value of each of the twelve variables (things like wind speed, air pressure, humidity, temperature, etc.) in his model. To save paper, space and time, and to make the results more palatable, rather than spitting out the whole length of the number that the computer was working with, it printed a truncated representation rounded to three decimal places – accurate enough to see how the system was evolving over time, but less accurate than the six decimal places the computer was working with internally. A value that the computer stored as 24.120034, for example, would print simply as 24.120.

Lorenz decided to repeat one of his simulations but didn't want to go all the way back to the start, so instead used the truncated print-ed-out values of the variables from halfway through his previous run as initial conditions for this new repeat. When he checked on the final results of this rerun, he was dismayed to find that, although the predicted weather patterns stayed similar for a while, by the end of the simulation, his two repeats were predicting totally different weather. The difference of less than one in a thousand in the values of his initial conditions were causing the system to diverge after relatively short periods of time.

Lorenz puzzled about this result for a while, suspecting problems with his computer and eventually simplifying his model down to just three variables to understand it better,[192] but the problem didn't go away. After repeating his experiment multiple times on different machines, he eventually concluded, against the prevailing opinion at the time (which was that small differences in the initial conditions shouldn't make much difference to the final output of the model) that this sensitive dependence on initial conditions was an inherent property of the system – a hallmark of chaotic systems.

He shared his discoveries in a paper[193] entitled 'Does the flap of a butterfly's wings in Brazil set off a tornado in Texas?' which he presented to the American Association for the Advancement of Science in December 1972. The purpose of his poetic question was to illustrate that without perfect knowledge of the initial conditions, we can't expect to be able to predict future weather patterns. Indeed, far enough into the future, he suggested, we may see two distinct predictions from near-identical initial conditions – one in which a tornado is forecast and another in which it isn't.

The *butterfly effect* was a powerful analogy and doubly fitting. The trajectories of Lorenz's simplified atmospheric model, which he used to emphasise the chaotic behaviours that could be exhibited by deterministic systems, look (with a little imagination and when viewed from the right angle) like the spread wings of a butterfly in flight (see Figure 9-2). The butterfly effect has become one of the most famous ideas in popular science, bleeding through heavily into popular culture. I counted well over a hundred songs on Spotify with the name '(The) Butterfly Effect' before I got bored and gave up. There are over twenty albums with the same name and at least two bands. IMDB references at least a hundred films or TV episodes with the phrase in the title.

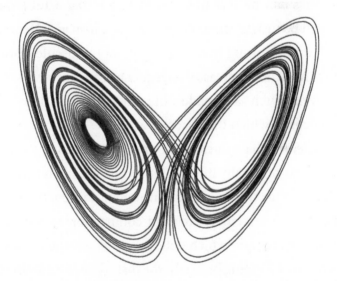

Figure 9-2: Edward Lorenz's chaotic butterfly. The trajectories of the simplified model of the atmosphere, which Lorenz used to characterise chaos, map out a shape which resembles the wings of a butterfly, when viewed from this angle.

Unfortunately, when translated from scientific illustration to pop-culture trope, the nuance of the analogy is often misappropriated or misrepresented. Robert Redford's character, the mathematician and gambler Jack Weil, in the 1990 movie *Havana*, claims, 'A butterfly can flutter its wings over a flower in China and cause a hurricane in the Caribbean. *I believe it. They can even calculate the odds.*' This is almost the opposite of what Lorenz was intending to convey with his provocative butterfly question: that the sensitivity to the initial conditions would make these types of systems practically unpredictable beyond a certain time horizon. Certainly, it would be impossible to attribute the formation of a hurricane directly to any one flap of a butterfly's wings or to any particular butterfly.

As Lorenz recognised in his original paper: 'If the flap of a butterfly's wings can be instrumental in generating a tornado, it can equally well be instrumental in preventing a tornado'. He went on to clarify: 'I am proposing that over the years minuscule disturbances neither increase nor decrease the frequency of occurrences of various weather events such as tornados; the most they may do is to modify the sequences in which they occur.'

Some scientists have questioned whether the analogy makes sense, even as Lorenz originally intended it. While each flap of a butterfly's wings does change the air pressure around it, this fluctuation quickly dissipates and is incredibly small in comparison to the large-scale changes in air pressure which determine the weather. Within a few centimetres of a flapping butterfly, the disturbance it causes will have been dissipated by the surrounding air molecules, making it difficult to imagine how such minute changes could be amplified fast enough to manifestly change the forecasted weather to the degree required to trigger or avert a tornado. More likely, the doubters suggest, our crude representations of the earth's atmosphere – glossing over some important facets of the physics governing the weather – have a bigger impact than the flapping of a single, or even all the world's butterflies.

Off limits

Whatever the reason – model inaccuracy or genuine chaotic behaviour – there are limits to our predictive capabilities. Maths can only take us so far. Even with the most powerful computational resources in the world thrown at the problem, there are fundamental restrictions on our ability to foretell the future: for how long, to what degree of accuracy and about what. Chaos puts fundamental limits on how far we can peer into the future – or indeed, how far we can look back into the past. The ubiquity of uncertainty and chaos mean we shouldn't try to make definitive predictions too far off into the future. And if we do cast our predictive nets a long way forward in time, we should be careful about how we interpret their haul.

Instead, the predictions we make, as with the predictions we might hope to read, should come with corresponding caveats about our degree of confidence in them. Although critics might argue that weather forecasters are just covering their backs when they provide the percentage chance that it will rain tomorrow, they are, in fact, doing something quite different. By admitting a degree of doubt in their predictions, forecasters provide more useful information than a blanket prediction of rain or sunshine. We should be wary of anyone who tells us they know what will happen in the future with absolute certainty. If we follow those purveyors of the 'sure thing' or the 'dead cert', who believe in their systems too much and don't admit to any degree of uncertainty, we are destined to be surprised when things don't turn out as expected.

Understanding our predictive limitations – the situations in which we cannot make good predictions – is of vital importance. It is just as valuable to know when we can't determine what will happen in the future as when we can. Misplaced confidence in a bad prediction is almost always worse than the circumspection which accompanies having no forecast at all. Arguably, understanding the science of how to make predictions is of less importance than having studied the art of knowing when not to.

EPILOGUE

We have been trying to predict the future for millennia, with varying degrees of success. In antiquity, the art of making seemingly correct predictions was largely based either on experience – the sun has come up every day we can remember, so we predict it will come up again tomorrow – or manipulation. Soothsayers would employ a combination of the broad range of tools which, we found in Chapter 1, are exploited by modern-day psychics – the Forer effect, the rainbow ruse, shotgunning, vanishing negatives and fishing expeditions – to score hits. Combining these techniques with knowledge of our biases – confirmation bias, hindsight bias, motivated reasoning and coincidence bias – made (and still make) us more likely to accept their speculative guesswork and to forgive or forget their missteps.

More recently, the era of science – and perhaps most prominently, mathematics – has ushered in a new age of prediction. We were able to move beyond simply relying on experience – predicting weather by looking back in the history books at what had happened historically on a given day. Instead, we were able to generalise the phenomena we experience – to hypothesise general laws – which allowed us to predict reliably about scenarios we had not yet experienced. We began to be able to predict in the face of both aleatoric uncertainty ('Where will the planets be a hundred years from now?') and epistemic uncertainty ('What are the missing elements in the periodic table?').

This practice of generalising experiences, making hypotheses and subsequently testing them, is the bedrock of science. Making predictions that were later falsified or validated by experiments was crucial to driving knowledge forwards. The concept of predicting in the face of epistemic uncertainty, in order to uncover the secrets of the universe, encapsulates the paradigm of scientific investigation – reasoning about situations which are (we hope) knowable, but as yet unknown.

As our scientific knowledge and practices improved, we were able to make better, more reliable and more general predictions. Correspondingly, as we learned more about how to predict, we were able to answer more of the scientific questions that arose – a symbiotic positive feedback loop. In this sense, even to this day, the best scientists relish making predictions that may later turn out to be incorrect. The very act of getting something wrong teaches us something about the world – even if it is perhaps not what we were expecting.

Science also provided us with the tools to expose those unscrupulous characters who were out to exploit us with their unfounded rhetoric. With controlled trials we were able to debunk the snake-oil salesmen making fantastic, but unsupported, claims about their miracle cures. We were also able to shine an objective light on fraudulent soothsayers, discovering that we could see right through them – their success rates being no better than those dictated by chance alone (and sometimes worse). Indeed, the mathematics of probability, by demonstrating how often seemingly unlikely things can and do happen, stayed our hands when grasping at unwarranted conclusions – inferring causation when all that existed was correlation or coincidence.

As comprehensive as the scientific revolution appeared, in the nineteenth and twentieth centuries, we began to discover hard limits to how well we could ever hope to predict, even with the most accurate scientific and mathematical tools available. Chaos dictates that even with incredibly detailed models of complex nonlinear phenomena, our predictions will still eventually diverge from reality. Some parts of the future lie hidden to us beyond a horizon in time. The randomness inherent in many of the systems we would like to make predictions about – from sports events to hurricanes – means we can never predict what will happen in the future with absolute certainty. The best we can hope is to give a range of probable scenarios accompanied by a numerical value expressing our confidence in each one.

For some phenomena, like earthquakes, we can't even manage this. The distribution of frequencies with which future earthquakes occur may be well characterised at an aggregate level, but that doesn't help us predict when the next one will occur and how devastating it might be.

That said, understanding the frequency of different-magnitude earthquakes in different places helps us to prepare and prioritise resources, even if we don't know exactly when the next one will hit. In the UK, we don't routinely prepare for earthquakes because these distributions indicate that the chances of experiencing large-magnitude earthquakes in this country are extremely low. In contrast, Japan routinely spends upwards of 3 per cent of its annual budget on disaster-risk management.

Since strengthening disaster preparedness began in Japan in the late 1950s, average annual deaths have been reduced from the thousands to the low hundreds. Japan even has a national 'Disaster Prevention Day' on 1 September each year – the anniversary of the 1923 Great Kantō earthquake. The day acts as a catalyst for people to take part in disaster drills, as well as an opportunity for the government to disseminate disaster-preparedness advice to the public.

Every plan we make represents a wager against the world's uncertainties. Preparation is no different. The degree to which we prepare represents the trade-off between what we are willing to sacrifice now to hedge our bets against the vagaries of the future. Buying critical-illness cover is a routine way to tacitly express our degree of confidence in the likelihood of our own ill health. Building a nuclear bunker or making plans to live entirely 'off grid' represent more extreme hedges against, perhaps less likely, future scenarios.

It's important to remember that choosing not to prepare for any given eventuality is also an implicit prediction. At an individual level, choosing not to take out life insurance might be considered a bet against unexpected death. At a national scale, failing to stockpile personal protective equipment or build health-service capacity are the actions of a country implicitly betting against a pandemic. Fathoming the future is not just about the 'positive' predictions we explicitly formulate, but also the 'negative' ones we don't. The latter, by their absence, are often harder to spot, but their failure can be equally damaging.

Some people believe that the failure to detect and prevent the 9/11 terrorist attacks represents just such a failure of prediction. Other commentators believe, with good reason, that the attacks were either wholly or largely unpredictable. They suggest that being so far from the realm of what we had previously experienced, we lacked the relevant data to feed into predictive models that might have credibly alerted the world to the possibility of a terrorist attack of this magnitude on American soil.

Looking back on the lead-up to 9/11 in retrospect, it might not seem that way. Many, in hindsight, have suggested measures that might have prevented this terrorist atrocity. Airports might have improved their security screening, cockpit doors might have been locked, the threat of a plane hijacking might have been taken more seriously. But it seems unreasonable to retrospectively blame our past selves for not prospectively having implemented these precautions. Although it is hard to remember when viewed through the lens of hindsight, the risk of terrorism was simply not as heightened before 9/11 as it became after. Had anyone made such suggestions and been strident enough to convince authorities to take these measures – banning all but the smallest amount of liquid on an aircraft, making people remove their shoes when going through airport security – you can be sure there would have been strong and vocal political factions advocating against these 'unnecessary, freedom-restricting' measures. Indeed, had 9/11 been averted, making the prediction self-defeating, these hypothetical security measures may have eventually loosened, facilitating an environment for a similarly catastrophic attack to occur.

To critique pre-9/11 counterterrorism initiatives solely on the basis that the attacks were allowed to occur is also to suffer from survivorship bias – forgetting the number of foiled terrorist plots we never get to hear about. In order to even try to foresee these sorts of unusual events, we need to engage in counterfactual thinking – looking back at the past and imagining events had taken a different course, asking the 'what-if' and 'if-only' questions that allow us to envisage alternative presents and alternative futures. It goes without saying that to imagine events playing out contrary to our experience of them in reality is difficult to achieve.

This is a problem I encountered when planning a stunt for this book in January 2020. I set out to pull a similar trick to the one I described in Chapter 3 – betting on both sides of each of five (what could essentially be considered) two-horse races. My bets on this occasion were to be on Liverpool and Man City to win the English Premier League, as well as both Oxford and Cambridge to win the 2020 men's boat race (both bets which came up again in the 2022 accumulator bets I described earlier in the book), Tyson Fury and Deontay Wilder to win their boxing rematch, Novak Djokovic and Dominic Thiem to win the Australian Open men's singles title (once it was decided that the two men would contest the final), and the Democratic and Republican candidates to win the US presidential election (Joe Biden had not yet been selected as the Democratic candidate). I thought I had covered all the angles – that there was no way I could lose this bet, because I would bet on all the possible outcomes. What I forgot to consider was the wider context – the fact that the outcomes (or more precisely that there would even be outcomes) of these events were contingent on the world continuing to behave, more or less, as I had always experienced it. This, it turns out in hindsight, was a bad assumption, but perhaps a forgivable one.

As it transpired, I was too disorganised to place the necessary thirty-two bets in time. For once, my disorganisation paid dividends. Novak Djokovic did go on to win the men's singles title at the Australian Open in early February of 2020 and Tyson Fury defeated Deontay Wilder in their first rematch that same month. Liverpool eventually won the English Premier League by a huge margin and Democratic candidate Joe Biden went on to win a hotly disputed US presidential election. But the 2020 boat race never took place. The Covid pandemic struck in the early months of 2020. The boat race was due to be run on 29 March 2020, by which point the UK was a week into our first and most-stringent lockdown. People were only allowed out of their homes for medical emergencies, essential shopping, travelling to work or one bout of solo exercise. Certainly, the boat race couldn't have gone ahead on its planned date. Had I placed the accumulator bets, the emergence of SARS-CoV-2 as a global threat would have scuppered what I had considered to be a foolproof plan to convince the world I could predict the future with certainty. Even though I was sure my bet would be a dead cert, it didn't turn out that way in the end.

At the time I was planning that bet, I was also extremely confident that I would finish the first draft of this book by spring 2021 for publication early in 2022. As it transpired, several months of home-schooling my two kids and regular science communication commitments throughout the first two years of the pandemic meant that I eventually handed in my first draft over a year late. It just goes to show – even the best-laid plans can fail.

If we fail to carefully think through the potential consequences of our actions, our plans can go awry despite our best intentions. When Richard Gatling invented his rapid-fire hand-cranked machine gun in 1861, he genuinely thought it might save lives. Gatling had noticed that more soldiers were invalided away from the battlefield through illness than as a result of bullet wounds. Of his new invention he wrote:

> It occurred to me that if I could invent a machine – a gun – which could by rapidity of fire, enable one man to do as much battle duty as a hundred, that it would, to a great extent, supersede the necessity of large armies, and consequently, exposure to battle and disease [would] be greatly diminished.

At that point, at the beginning of the US Civil War, a highly skilled rifleman could fire five rounds a minute. The Gatling gun could fire 200 rounds a minute and required relatively little expertise. What his invention inadvertently achieved was to hugely increase the amount of 'battle duty' each army was able to perform. His gun backfired on him.

In 1883, Hiram Maxim patented the natural successor to Gatling's gun – the first automatic machine gun. It seemed that he too had high hopes that his invention might reduce war deaths, but for different reasons. When asked by the English scientist Havelock Ellis, 'Will this gun not make war more terrible?' he is alleged to have replied, 'No. It will make war impossible.' The Maxim gun was capable of firing over 600 rounds a minute. It was a killing machine. So destructive did Maxim apparently deem his invention that he believed, from a game-theoretic standpoint, it would be too devastating for both sides for military leaders to release such awesome power on the battlefield. The *New York Times* agreed with Maxim, suggesting that leaders would instead be forced to negotiate their way to peaceful settlements and characterising machine guns as 'peace-producing and peace-retaining terrors'.

Sadly, Maxim and the *New York Times*' cost–benefit analyses were way off. They underestimated man's inhumanity to man and overestimated the cost attached to a human life by the people in whose hands the power of life and death resided. In the First World War, machine guns are thought to have killed hundreds of thousands of men. The first day of the Battle of the Somme alone saw over 20,000 British soldiers mown down – the vast majority by the German equivalent of the Maxim gun. Weapons so destructive that they might never be used in anger would eventually be created, as we saw in Chapter 5, but the cost of their use – the potential to wipe out the entire human race – was so much higher than that of Maxim's gun. Even nuclear weapons have not proved terrifying enough to prevent conventional warfare.

Usually when a prediction goes wrong, there is something we can learn from the experience to help us the next time we are facing a similar situation. If there is just one lesson to take away from this book, it's that when plans do go wrong, we should evaluate why and try to learn how to prevent the same mistake in the future.

If we unthinkingly make a linear prediction, which turns out to be incorrect, we should question what it was about the process that made it fail. Was there an underlying positive feedback loop that snowballed out of control, growing faster than expected? Conversely, was there a hidden negative feedback loop which damped down an expected excursion from the status quo?

Thankfully, many mistakes have already been made for us, saving us the ignominy of making them again. But if we fail to learn the lessons of incorrect prediction from history, then we may end up like the hapless British occupiers of Chapter 8, repeatedly offering perverse incentives, or the scientists of Chapter 2 who, suffering from motivated reasoning, read too much into their data. When we consider predictions from the past that have seemed to come true, we should be careful that we are not subconsciously employing postdiction – fitting the prediction to the facts only after the event. Conversely, when we laugh at the seeming naïvety of failed predictions, we should be careful to check that we are not suffering from hindsight bias – making the events which transpired seem more predictable than they actually were.

When making our own inferences, we need to be aware of the inherent biases which can trip us up. Have we inferred too much from a perceived connection which might transpire to be nothing more than a coincidence? Have we spotted a 'hidden' pattern in the noise, which, it turns out, is just noise itself? Have we asked ourselves whether we are really seeing the whole picture, or whether we are just looking at the target that someone has drawn around the data and that tells the story they want us to hear?

If we can trust ourselves to change our opinions in the face of new evidence, no matter how wedded we were to the views we espoused in the past; if we can talk ourselves out of ever being 100 per cent certain on anything, so that there is room for new information to change our minds; and if we can rely on ourselves to use each piece of relevant data to update our opinions, then slowly, but surely we can hope to learn to expect what was previously unexpected.

ACKNOWLEDGMENTS

I got about halfway through writing this book before the Covid pandemic hit. My editors, Katy Follain and Nina Sandelson, and everyone at Quercus were kind enough to allow me to take a sabbatical from writing to focus on communicating the mathematics behind the pandemic to the public. I was fortunate enough to be able to share the importance of mathematical modelling and statistics across platforms, from national newspapers to television news programmes to international radio current-affairs shows. I felt we were able to provide a very direct answer to the age-old question, beloved by schoolchildren in maths classrooms everywhere, 'When am I ever going to need to use this?'

Some of the many stories I covered in the news made their way into this book – examples of failures of prediction, victims of linearity bias and of normalcy bias, people who failed to understand exponential growth. But while the book may have benefitted in one way from my experiences, it certainly suffered in others – most notably, for about twelve months I didn't write a single word. The irony of writing a book about prediction and not being able to foresee its derailment by an event of global magnitude wasn't lost on me.

During the acute phase of the pandemic, I witnessed some fantastic science communication, and made many excellent and long-lasting connections, not least my colleagues in Independent SAGE – an independent scientific advisory group sharing science relating to the pandemic directly with the general public. I have learned so much from my colleagues on Independent SAGE and their support has meant a great deal to me.

As ever, once I eventually got back to writing, the assistance of my dad, Tim, and step-mum, Mary, has been incredibly important. They have read everything in the book at least once over and provided me with helpful comments and corrections. Similarly, my two former PhD colleagues, Aaron Smith and Gabriel Rosser, have been brilliant fact-checkers and have known how to pose the right challenges in the right way.

I also want to thank my employers, the University of Bath, who have been encouraging in my science communication efforts and, in particular, my colleague Jon Dawes who sense-checked the parts of the book that were concerned with chaos.

As ever, my agent, Chris Wellbelove, and my editors, Katy Follain and Nina Sandelson, have stuck by my side and fought my corner. Your feedback and support have been invaluable. I look forward to finding out what the future will bring for our collaboration.

I owe a debt of gratitude to all the people whom I contacted when writing this book and who so kindly agreed to share their stories with me. It is your experiences which are the substance of this book and I appreciate the effort you all went to to share them with me.

Finally, my biggest thanks must go to my family for once again encouraging me and putting up with me when I have been consumed with writing. In particular, my wife, Caz, is my biggest supporter. My children, Will and Emmie, are my most patient and faithful advocates. This book is for them. You are the future.

REFERENCES

Introduction

1   Page 3 'in the prestigious journal *Science*'
    Rasool, S. I., & Schneider, S. H. (1971). Atmospheric carbon dioxide and aero-
    sols: Effects of large increases on global climate. *Science, 173*(3992), 138–41.
    https://doi.org/10.1126/science.173.3992.138

2   Page 4 'the species introduced to control a pest which ultimately became a
    scourge themselves'
    Easteal, S. (1981). The history of introductions of Bufo marinus (Amphibia:
    Anura); a natural experiment in evolution. *Biological Journal of the Linnean
    Society, 16*(2), 93–113. https://doi.org/10.1111/J.1095-8312.1981.TB01645.X

3   Page 6 'the fusiform face area – responsible for recognising and remembering
    them'
    Kanwisher, N., McDermott, J., & Chun, M. M. (1997). The fusiform face
    area: A module in human extrastriate cortex specialized for face perception.
    Journal of Neuroscience, 17(11), 4302–11. https://doi.org/10.1523/jneurosci.17-
    11-04302.1997

4   Page 8 'economists gather data from the more distant past in order to make
    "nowcasts"'
    Lahiri, K., & Monokroussos, G. (2013). Nowcasting US GDP: The role of ISM

business surveys. *International Journal of Forecasting*, 29(4), 644–58. https://doi.org/10.1016/j.ijforecast.2012.02.010

5   Page 8 'health researchers feed social-media data into nowcasting models to detect flu epidemics'
Lampos, V., & Cristianini, N. (2012). Nowcasting events from the social web with statistical learning. *ACM Transactions on Intelligent Systems and Technology*, 3(4). https://doi.org/10.1145/2337542.2337557

6   Page 9 'The Ancient Egyptians believed the earth was a flat disc'
Frankfort, H., Frankfort, H. A., Wilson, J. A., Jacobsen, T., & Irwin, W. A. (1948). The Intellectual Adventure of Ancient Man: An Essay on Speculative Thought in the Ancient near East. *The Journal of Religion*, 28(3), 210–13. https://doi.org/10.1086/483727

7   Page 14 'the location of missing planets'
Smart, W. M. (1946). John Couch Adams and the discovery of Neptune. *Nature*, 158(4019), 648–52. https://doi.org/10.1038/158648a0

8   Page 14 'the existence of radio waves'
Maxwell, J. C. (1865). A dynamical theory of the electromagnetic field. *Philosophical Transactions of the Royal Society of London*, 155, 459–512. https://doi.org/10.1098/rstl.1865.0008

9   Page 14 'the pitter-patter of a dripping tap'
Cahalan, R. F., Leidecker, H., & Cahalan, G. D. (1990). Chaotic Rhythms of a Dripping Faucet. *Computers in Physics*, 4(4), 368. https://doi.org/10.1063/1.4822928

10   Page 14 'fluctuations of animal populations'
May, R. M. (1987). Chaos and the dynamics of biological populations. *Proceedings of The Royal Society of London, Series A: Mathematical and Physical Sciences*, 413(1844), 27–44. https://doi.org/10.1515/9781400860197.27

Chapter 1

11   Page 28 'A 2005 survey by the US polling company Gallup'
Moore, D. W. (2005). *Three in Four Americans Believe in Paranormal: Little Change From Similar Results in 2001*. Gallup Poll News Service. http://www.gallup.com/poll/16915/Three-Four-Americans-Believe-Paranormal.aspx

12   Page 29 '*Barnum statements*'

Dickson, D. H., & Kelly, I. W. (1985). The 'Barnum Effect' in Personality Assessment: A Review of the Literature. *Psychological Reports, 57*(2), 367–382. https://doi.org/10.2466/pr0.1985.57.2.367

13   Page 30 '*Forer effect*'

Howard, J. (2019). Forer Effect. *Cognitive Errors and Diagnostic Mistakes,* 139–44. https://doi.org/10.1007/978-3-319-93224-8_9

14   Page 33 '*Pollyanna principle*'

Matlin, M. W., and Stang, D. J. (1978). *The Pollyanna principle: Selectivity in language, memory, and thought.* Schenkman Publishing Company.

15   Page 33 'why compliments make us feel good'

Izuma, K., Saito, D. N., & Sadato, N. (2008). Processing of Social and Monetary Rewards in the Human Striatum. *Neuron, 58*(2), 284–94. https://doi.org/10.1016/j.neuron.2008.03.020

16   Page 35 'Carl Jung first introduced the concept'

Jung, C. G. (1952). *Synchronicity: an acausal connecting principle.* Princeton University Press.

17   Page 36 'He carefully designed an experiment'

Ono, K. (1987). Superstitious behavior in humans. *Journal of the Experimental Analysis of Behavior, 47*(3), 261. https://doi.org/10.1901/JEAB.1987.47-261

18   Page 37 'conducted an experiment with three- to six-year-old children'

Wagner, G. A., & Morris, E. K. (1987). 'Superstitious' Behavior in Children. *The Psychological Record, 37*(4), 471–88. https://doi.org/10.1007/bf03394994

19   Page 39 'the *frequency illusion*'

Zwicky, A. M. (2006). Why are we so illuded? https://web.stanford.edu/~-zwicky/LSA07illude.abst.pdf

20   Page 45 'von Restorff discovered in her experiments to remember an unusual item on a list of otherwise similar objects'

Von Restorff, H. (1933). Über die Wirkung von Bereichsbildungen im Spurenfeld. *Psychologische Forschung, 18*(1), 299–342. https://doi.org/10.1007/BF02409636

Chapter 2

21   Page 66 'Wegener proposed the only theory that reconciled his observations'
      Wegener, A. (1912). Die Herausbildung der Grossformen der Erdrinde
      (Kontinente und Ozeane), auf geophysikalischer Grundlage (The uprising of
      large features of earth's crust (Continents and Oceans) on geophysical basis).
      Petermanns Geographische Mitteilungen, 63, 185–195.

22   Page 66 'When he published his theory in 1915'
      Wegener, A. (1929). *Die entstehung der kontinente und ozeane (The origin of
      continents and oceans)* (4th ed.). Braunschweig: Friedrich Vieweg & Sohn Akt.
      Ges.

23   Page 66 'By the 1960s, however, the theory of plate tectonics'
      Le Pichon, X. (1968). Sea-floor spreading and continental drift. *Journal
      of Geophysical Research, 73*(12), 3661–3697. https://doi.org/10.1029/
      jb073i012p03661

24   Page 66 'John Dalton had recently measured were roughly whole number mul-
      tiples of the atomic weight of hydrogen'
      Dalton, J. (1806). III. On the absorption of gases by water and other liquids . *The
      Philosophical Magazine, 24*(93). https://doi.org/10.1080/14786440608563325

25   Page 66 'This led him to suggest that atoms of other elements would be amal-
      gamations of various numbers of hydrogen atoms'
      Prout, W. (1815). On the relation between the specific gravities of bodies in their
      gaseous state and the weights of their atoms. *Annals of Philosophy, 6,* 321–330.
      https://web.lemoyne.edu/~giunta/PROUT.HTML

26   Page 66 'Chlorine presented a particular problem'
      Harkins, W. D. (1925). The Separation of Chlorine into Isotopes (Isotopic
      Elements) and the Whole Number Rule for Atomic Weights. *Proceedings of
      the National Academy of Sciences, 11*(10), 624–628. https://doi.org/10.1073/
      pnas.11.10.624

27   Page 67 'Ernest Rutherford fired alpha particles at nitrogen atoms to displace
      hydrogen nuclei'
      Rutherford, E. (1919). LIV. Collision of α particles with light atoms . IV.
      An anomalous effect in nitrogen. *The London, Edinburgh, and Dublin*

Philosophical Magazine and Journal of Science, 37(222), 581–587. https://doi. org/10.1080/14786440608635919

28 Page 68 'in what he called recapitulation theory'
Mayr, E. (1982). *The Growth of Biological Thought: Diversity, Evolution, and Inheritance*, 974. Harvard University Press.

29 Page 68 'it was discovered in 1827 that human embryos really do have slits'
Rathke, M. H. (1828). Über das Dasein von Kiemenandeutungen bei menschlichen Embryonen (On the existence of gill slits in human embryos). *Isis von Oken, 21*, 108–9.

30 Page 69 'the idea of common descent started to take hold'
Darwin, C. (1859). *On the origin of species by means of natural selection, or the preservation of favoured races in the struggle for life. On the origin of species by means of natural selection, or the preservation of favoured races in the struggle for life*. John Murray.

31 Page 71 'Behavioural scientists have shown'
Huber, J., Payne, J. W., & Puto, C. (1982). Adding Asymmetrically Dominated Alternatives: Violations of Regularity and the Similarity Hypothesis. *Journal of Consumer Research, 9*(1), 90. https://doi.org/10.1086/208899

32 Page 72 'educational psychologists have found'
Attali, Y., & Bar-Hillel, M. (2003). Guess where: The position of correct answers in multiple-choice test items as a psychometric variable. *Journal of Educational Measurement, 40*(2), 109–28. https://doi.org/10.1111/j.1745-3984.2003. tb01099.x

33 Page 72 'choosing items on a shelf or options from a computer drop-down menu'
Bar-Hillel, M. (2015). Position effects in choice from simultaneous displays: A conundrum solved. *Perspectives on Psychological Science, 10*(4), 419–33. https:// doi.org/10.1177/1745691615588092

34 Page 72 'middle cubicle is up to 50 per cent more likely to be chosen than an outer one'
Christenfeld, N. (1995). Choices from Identical Options. *Psychological Science, 6*(1), 50–55. https://doi.org/10.1111/j.1467-9280.1995.tb00304.x

35 Page 73 'some of my recent research'
Gavagnin, E., Owen, J. P., & Yates, C. A. (2018). Pair correlation functions for

identifying spatial correlation in discrete domains. *Physical Review E, 97*(6), 062104. https://doi.org/10.1103/PhysRevE.97.062104

36   Page 73 'better understand the beautiful striped patterns of zebrafish'
Owen, J. P., Kelsh, R. N., & Yates, C. A. (2020). A quantitative modelling approach to zebrafish pigment pattern formation. *ELife, 9*, 1–62. https://doi.org/10.7554/eLife.52998

37   Page 77 'a huge study from Sweden was published'
Feychting, M., & Alhbom, M. (1993). Magnetic fields and cancer in children residing near Swedish high-voltage powerlines. *American Journal of Epidemiology, 138*(7), 467–81. https://doi.org/10.1093/oxfordjournals.aje.a116881

38   Page 83 'remembering roughly two dream themes per week'
Goodenough, D. R. (1991). Dream recall: History and current status of the field. In *The mind in sleep: Psychology and psychophysiology* (pp. 143–71). John Wiley & Sons.

39   Page 83 'most commonly remembered dream themes'
Morewedge, C. K., & Norton, M. I. (2009). When Dreaming Is Believing: The (Motivated) Interpretation of Dreams. *Journal of Personality and Social Psychology, 96*(2), 249–64. https://doi.org/10.1037/a0013264

Chapter 3

40   Page 112 'one study on AIDS and sexual behaviour undertaken in France in 1992'
Spira, A., Bajos, N., Béjin, A., Beltzer, N., Bozon, M., Ducot, B., . . . Touzard, H. (1992). AIDS and sexual behaviour in France. *Nature, 360*, 407–409. https://doi.org/10.1038/360407a0

41   Page 113 'research in which a treatment has no discernible effect'
Dickersin, K., Chan, S., Chalmersx, T. C., Sacks, H. S., & Smith, H. (1987). Publication bias and clinical trials. *Controlled Clinical Trials, 8*(4), 343–53. https://doi.org/10.1016/0197-2456(87)90155-3

42   Page 113 'results can be distorted by publication bias'
Kicinski, M., Springate, D. A., & Kontopantelis, E. (2015). Publication bias in

meta-analyses from the Cochrane Database of Systematic Reviews. *Statistics in Medicine*, 34(20), 2781–93. https://doi.org/10.1002/sim.6525

43   Page 114 'a study which looked at cats taken to vets after falling from high-rise buildings in the late 1980s'
Whitney, W. O., & Mehlhaff, C. J. (1987). High-rise syndrome in cats. *Journal of the American Veterinary Medical Association*, 191(11), 1399–403. https://europepmc.org/article/med/3692980

44   Page 118 'dopamine rush'
Rutledge, R. B., Skandali, N., Dayan, P., & Dolan, R. J. (2014). A computational and neural model of momentary subjective well-being. *Proceedings of the National Academy of Sciences of the United States of America*, 111(33), 12252–7. https://doi.org/10.1073/pnas.1407535111

45   Page 118 'keeps them coming back, win or lose'
Narayanan, S., & Manchanda, P. (2012). An empirical analysis of individual level casino gambling behavior. *Quantitative Marketing and Economics*, 10(1), 27–62. https://doi.org/10.1007/s11129-011-9110-7

46   Page 119 'With players standing to lose 55 per cent of their stake'
Cox, S. J., Daniell, G. J., & Nicole, D. A. (1998). Using Maximum Entropy to Double One's Expected Winnings in the UK National Lottery. *Journal of the Royal Statistical Society: Series D (The Statistician)*, 47(4), 629–641. https://doi.org/10.1111/1467-9884.00160

47   Page 123 'when people choose their lottery numbers'
Ibid.

48   Page 123 'Actively taking advantage of other people's randomness biases'
Ibid.

49   Page 126 'the team of neuropsychologists conducting the study'
Schulz, M.-A., Schmalbach, B., Brugger, P., & Witt, K. (2012). Analysing Humanly Generated Random Number Sequences: A Pattern-Based Approach. *PLoS ONE*, 7(7), e41531. https://doi.org/10.1371/journal.pone.0041531

50   Page 129 'analysed thousands of commuter journeys before, during and after the strike'
Larcom, S., Rauch, F., & Willems, T. (2017). The Benefits of Forced Experimentation: Striking Evidence from the London Underground Network. *The Quarterly Journal of Economics*, 132(4), 2019–55. https://doi.org/10.1093/qje/qjx020

51 Page 130 'more recent theorists have suggested that increased choice can induce a range of anxieties in consumers'
Schwartz, B. (2004). *The Paradox of Choice: Why More is Less.* Ecco.

52 Page 131 'researchers from Columbia and Stanford Universities set out to explore exactly this hypothesis'
Iyengar, S. S., & Lepper, M. R. (2000). When choice is demotivating: Can one desire too much of a good thing? *Journal of Personality and Social Psychology*, 79(6), 995–1006. https://doi.org/10.1037/0022-3514.79.6.995

53 Page 133 'a randomly dictated decision prompt can help to deal with the information overload that often precipitates analysis paralysis'
Douneva, M., Jaffé, M. E., & Greifeneder, R. (2019). Toss and turn or toss and stop? A coin flip reduces the need for information in decision-making. *Journal of Experimental Social Psychology*, 83, 132–41. https://doi.org/10.1016/j.jesp.2019.04.003

Chapter 4

54 Page 141 'the distribution of patients' blood pressure'
Pater, C. (2005). The blood pressure 'uncertainty range' – A pragmatic approach to overcome current diagnostic uncertainties (II). In *Current Controlled Trials in Cardiovascular Medicine*, 6(1), 5. BioMed Central. https://doi.org/10.1186/1468-6708-6-5

55 Page 141 'human male and female heights'
Jelenkovic, A., Sund, R., Hur, Y. M., Yokoyama, Y., Hjelmborg, J. V. B., Möller, S., Honda, C., Magnusson, P. K. E., Pedersen, N. L., Ooki, S., Aaltonen, S., Stazi, M. A., Fagnani, C., D'Ippolito, C., Freitas, D. L., Maia, J. A., Ji, F., Ning, F., Pang, Z., . . . Silventoinen, K. (2016). Genetic and environmental influences on height from infancy to early adulthood: An individual-based pooled analysis of 45 twin cohorts. *Scientific Reports*, 6(1), 1–13. https://doi.org/10.1038/srep28496

56 Page 141 'the first digits of the data will display a very specific pattern – Benford's law'
Hill, T. P. (1995). A Statistical Derivation of the Significant-Digit Law. *Statistical Science*, 10(4), 354–363. https://doi.org/10.1214/ss/1177009869

57 Page 142 'might have alerted auditors to the fraud sooner'
Nigrini, M. J. (2005). An Assessment of the Change in the Incidence of Earnings Management Around the Enron-Andersen Episode. In *Review of Accounting and Finance*, 4(1), 92-110. Emerald Group Publishing Limited. https://doi.org/10.1108/eb043420

58 Page 143 'the Greek figures were found to have the highest divergence away from Benford's distribution, suggesting that they had been manipulated'
Rauch, B., Göttsche, M., Engel, S., & Brähler, G. (2011). Fact and Fiction in EU-Governmental Economic Data. *German Economic Review*, 12(3), 243–55. https://doi.org/10.1111/j.1468-0475.2011.00542.x

59 Page 143 'presidential-election results in Iran'
Roukema, B. F. (2014). A first-digit anomaly in the 2009 Iranian presidential election. *Journal of Applied Statistics*, 41(1), 164–99. https://doi.org/10.1080/02664763.2013.838664

60 Page 143 'fraudulent reporting in scientific research'
Horton, J., Krishna Kumar, D., & Wood, A. (2020). Detecting academic fraud using Benford law: The case of Professor James Hunton. *Research Policy*, 49(8), 104084. https://doi.org/10.1016/j.respol.2020.104084

61 Page 143 'day-to-day accounts auditing'
Nigrini, M. J. (1999). I've got your number. *Journal of Accountancy*, 187(5), 79–83.

62 Page 146 'for many other languages, even the artificial language Esperanto'
Manaris, B., Pellicoro, L., Pothering, G., & Hodges, H. (2006). Investigating Esperanto's statistical proportions relative to other languages using neural networks and Zipf's law. *Proceedings of the IASTED International Conference on Artificial Intelligence and Applications, AIA 2006*.

63 Page 147 'the number of papers written by scientists'
Lotka, A. (1926). The frequency distribution of scientific productivity. *Journal of the Washington Academy of Sciences*, 16(12), 317–23.

64 Page 147 'the population size of settlements'
Gabaix, X. (1999). Zipf's Law for Cities: An Explanation. *The Quarterly Journal of Economics*, 114(3), 739–67. https://doi.org/10.1162/003355399556133

65 Page 147 'immune-related amino-acid sequence lengths'
Mora, T., Walczak, A. M., Bialek, W., & Callan, C. G. (2010). Maximum entropy

models for antibody diversity. *Proceedings of the National Academy of Sciences of the United States of America, 107*(12), 5405–10. https://doi.org/10.1073/pnas.1001705107

66 Page 147 'the diameters of craters on the moon'
Neukum, G., & Ivanov, B. A. (1994). Crater Size Distributions and Impact Probabilities on Earth from Lunar, Terrestrial-planet, and Asteroid Cratering Data. In *Hazards due to comets and asteroids: Vol. Space Science Series*, 359–416.

67 Page 147 'the variation of species diversity with habitat area'
Martín, H. G., & Goldenfeld, N. (2006). On the origin and robustness of power-law species–area relationships in ecology. *Proceedings of the National Academy of Sciences of the United States of America, 103*(27), 10310–15. https://doi.org/10.1073/pnas.0510605103

68 Page 147 'frequency of the number of tornadoes per day in the United States'
Elsner, J. B., Jagger, T. H., Widen, H. M., & Chavas, D. R. (2014). Daily tornado frequency distributions in the United States. *Environmental Research Letters, 9*(2), 024018. https://doi.org/10.1088/1748-9326/9/2/024018

69 Page 147 'how the number of artists varies with the average price of their work'
Etro, F., & Stepanova, E. (2018). Power-laws in art. *Physica A: Statistical Mechanics and Its Applications, 506*, 217–20. https://doi.org/10.1016/j.physa.2018.04.057

70 Page 147 'the frequency of fatal conflicts varied with the number of people killed according to a power law'
Richardson, L. (1960). *Statistics of Deadly Quarrels*. Boxwood Press.

71 Page 148 'published by Charles Richter and Beno Gutenberg in 1956'
Gutenberg, B., & Richter, C. F. (2010). Magnitude and energy of earthquakes. *Annals of Geophysics, 53*(1), 7–12. https://doi.org/10.4401/ag-5590

72 Page 152 'This is the celebrated Gutenberg–Richter law'
Ibid.

73 Page 158 'before sharing the paper with the world'
Bayes, T., & Price, R. (1763). An essay towards solving a problem in the doctrine of chances. By the late Rev. Mr. Bayes, F. R. S. communicated by Mr. Price, in a letter to John Canton, A. M. F. R. S. *Philosophical Transactions of the Royal Society of London, 53*, 370–418. https://doi.org/10.1098/rstl.1763.0053

74 Page 158 'an age-old argument about sex ratios'
Laplace, P. (1778). Mémoire sur les probabilités. *Mémoires de l'Académie Royale des Sciences de Paris.*

75   Page 159 'employed it to help them hit their targets in the face of uncertain environmental conditions'

McGrayne, S. B. (2011). *The Theory That Would Not Die: How Bayes' Rule Cracked the Enigma Code, Hunted Down Russian Submarines, and Emerged Triumphant from Two Centuries of Controversy.* Yale University Press.

76   Page 159 'Alan Turing used it to help him crack Enigma'

Mardia, K. V., & Cooper, S. B. (2012). Alan Turing and Enigmatic Statistics. *Bulletin of the Brasilian Section of the International Society for Bayesian Analysis,* 5(2), 2–7.

77   Page 159 'a Russian submarine that had gone AWOL.'

Higgins, Chris. (2002). *Nuclear Submarine Disasters.* Chelsea House Publishers.

78   Page 159 'scientists used Bayes to help demonstrate the link between smoking and lung cancer'

Hill, A. B. (1950). Smoking and carcinoma of the lung preliminary report. *British Medical Journal,* 2(4682), 739–48. https://doi.org/10.1136/bmj.2.4682.739

79   Page 160 'from filtering out spam emails, from phishing attempts to pharmaceutical offers'

Sahami, M., Dumais, S., Heckerman, D., & Horvitz, E. (1998). A Bayesian approach to filtering junk e-mail. *Learning for Text Categorization: Papers from the AAAI Workshop.*

Chapter 5

80   Page 177 'historically, many cultures have considered suicide to be a rational response to ill health, dishonour or other forms of suffering'

Mayo, D. J. (1986). The Concept of Rational Suicide. *Journal of Medicine and Philosophy,* 11(2), 143–55. https://doi.org/10.1093/jmp/11.2.143

81   Page 178 'increases the chances that the progeny she sires will share his genetic material'

Schneider, J. M., Gilberg, S., Fromhage, L., & Uhl, G. (2006). Sexual conflict over copulation duration in a cannibalistic spider. *Animal Behaviour,* 71(4), 781–8. https://doi.org/10.1016/j.anbehav.2005.05.012

82   Page 178 'a trade-off worth making for the male orb-weaver'

Welke, K. W., & Schneider, J. M. (2010). Males of the orb-web spider *Argiope bruennichi* sacrifice themselves to unrelated females. *Biology Letters*, 6(5), 585–8. https://doi.org/10.1098/rsbl.2010.0214

83   Page 179 'the majority did not harbour psychological disorders and before their indoctrination by terrorist groups were relatively well-adjusted people'
Gunaratna, R. (2002). *Inside Al Qaeda: Global Network of Terror*. Columbia University Press.

84   Page 196 'penalty takers in two of Europe's top leagues chose randomly between kicking to the left, the right or down the middle'
Chiappori, P.-A., Levitt, S., & Groseclose, T. (2002). Testing Mixed-Strategy Equilibria When Players Are Heterogeneous: The Case of Penalty Kicks in Soccer. *American Economic Review*, 92(4), 1138–51. https://doi.org/10.1257/00028280260344678

85   Page 196 'In recent experiments into the impacts of emotional unpredictability'
Sinaceur, M., Adam, H., Van Kleef, G. A., & Galinsky, A. D. (2013). The advantages of being unpredictable: How emotional inconsistency extracts concessions in negotiation. *Journal of Experimental Social Psychology*, 49(3), 498–508. https://doi.org/10.1016/j.jesp.2013.01.007

86   Page 197 'leading them to make larger concessions and irresolute demands'
Ibid.

87   Page 203 'many of these will even die of their injuries'
Clutton-Brock, T. H., Albon, S. D., Gibson, R. M., & Guinness, F. E. (1979). The logical stag: Adaptive aspects of fighting in red deer (*Cervus elaphus* L.). *Animal Behaviour*, 27(PART 1), 211–25. https://doi.org/10.1016/0003-3472(79)90141-6

88   Page 204 'although he and his colleagues preferred to call it the "Sneaky Fucker" strategy'
Dawkins, R., & Krebs, J. R.   (1978). Animal Signals : Information or Manipulation? In J. R. Krebs & N. B. Davies (Eds.), *Behavioural Ecology: An Evolutionary Approach* (pp. 282–309). Blackwell Publishing.

89   Page 204 'A study of the mating habits of grey seals on Sable Island, off the coast of Canada, found that 36 per cent of females guarded by an alpha male were, in fact, fertilised by non-alpha males'
Ambs, S. M., Boness, D. J., Bowen, W. D., Perry, E. A., & Fleischer, R. C. (1999). Proximate factors associated with high levels of extraconsort fertilization

in polygynous grey seals. *Animal Behaviour, 58*(3), 527–35. https://doi.org/10.1006/anbe.1999.1201

90 Page 206 'voted against them in previous rounds'
Gonzalez, L. J., Castaneda, M., & Scott, F. (2019). Solving the simultaneous truel in *The Weakest Link*: Nash or revenge? *Journal of Behavioral and Experimental Economics* , *81*, 56–72. https://doi.org/10.1016/j.socec.2019.04.006

91 Page 207 'In 1968, a record 800,000 tonnes of fish were taken out of these fertile waters'
McCay, B. J., & Finlayson, A. C. (1995). The political ecology of crisis and institutional change: the case of the northern cod. In *Annual meeting of the American Anthropological Association, Washington, DC* (pp. 15–19).

92 Page 218 'made a huge difference to our consumption almost instantaneously'
Thomas, G. O., Sautkina, E., Poortinga, W., Wolstenholme, E., & Whitmarsh, L. (2019). The English Plastic Bag Charge Changed Behavior and Increased Support for Other Charges to Reduce Plastic Waste. *Frontiers in Psychology, 10* (Feb), 266. https://doi.org/10.3389/fpsyg.2019.00266

93 Page 219 'A zero ticket price can have the unintended consequence of lowering overall attendance, despite potentially increasing registrations'
Fan, X., Cai, F. C., & Bodenhausen, G. V. (2022). The boomerang effect of zero pricing: when and why a zero price is less effective than a low price for enhancing consumer demand. *Journal of the Academy of Marketing Science, 50*(3), 521–37. https://doi.org/10.1007/s11747-022-00842-1

Chapter 6

94 Page 228 'In the 1960s researchers undertook a series of experiments designed to understand human predictive behaviour'
Estes, W. K. (1961). A descriptive approach to the dynamics of choice behavior. *Behavioral Science, 6*(3), 177–184. https://doi.org/10.1002/bs.3830060302

95 Page 228 'pigeons employed a very different approach'
Hinson, J. M., & Staddon, J. E. R. (1983). Hill-climbing by pigeons. *Journal of the Experimental Analysis of Behavior, 39*(1), 25–47. https://doi.org/10.1901/jeab.1983.39-25

96 Page 234 'had the bubble not burst'

Huikari, S., Miettunen, J., & Korhonen, M. (2019). Economic crises and suicides between 1970 and 2011: Time trend study in 21 developed countries. *Journal of Epidemiology and Community Health*, 73(4), 311–16. https://doi.org/10.1136/jech-2018-210781

97 Page 245 'the function that is converged upon is the straight line indicating direct proportion'
Kalish, M. L., Griffiths, T. L., & Lewandowsky, S. (2007). Iterated learning: Intergenerational knowledge transmission reveals inductive biases. *Psychonomic Bulletin and Review*, 14(2), 288–94. https://doi.org/10.3758/BF03194066

98 Page 247 'our propensity to assume linearity is present long before we leave school'
De Bock, D., van Dooren, W., Janssens, D., & Verschaffel, L. (2002). Improper use of linear reasoning: An in-depth study of the nature and the irresistibility of secondary school students' errors. Educational Studies in Mathematics, 50(3), 311–34. https://doi.org/10.1023/A:1021205413749

99 Page 247 'the problems for which the true answer is available'
Van Dooren, W., de Bock, D., Hessels, A., Janssens, D., & Verschaffel, L. (2005). Not Everything Is Proportional: Effects of Age and Problem Type on Propensities for Overgeneralization. *Cognition and Instruction*, 23(1), 57–86. https://doi.org/10.1207/s1532690xci2301_3

100 Page 248 'many students were reluctant to abandon their original answers'
De Bock, D., van Dooren, W., Janssens, D., & Verschaffel, L. (2002). Improper use of linear reasoning: An in-depth study of the nature and the irresistibility of secondary school students' errors. *Educational Studies in Mathematics*, 50(3), 311–34. https://doi.org/10.1023/A:1021205413749

101 Page 248 'students believe a linear model is appropriate for every problem'
van Dooren, W., de Bock, D., Janssens, D., & Verschaffel, L. (2008). The linear imperative: An inventory and conceptual analysis of students' overuse of linearity. In *Journal for Research in Mathematics Education*, 39(3), 311-42.

102 Page 249 'fell prey to the deceptive linear logic'
De Bock, D., Verschaffel, L., & Janssens, D. (2002). The Effects of Different Problem Presentations and Formulations on the Illusion of Linearity in Secondary School Students. *Mathematical Thinking and Learning*, 4(1), 65–89. https://doi.org/10.1207/s15327833mtl0401_3

Chapter 7

103 Page 270 'avoid falling into the trap of underestimating exponential growth'
Levy, M., & Tasoff, J. (2016). Exponential-Growth Bias and Lifecycle Consumption. *Journal of the European Economic Association, 14*(3), 545–83. https://doi.org/10.1111/jeea.12149

104 Page 270 'people who have previously encountered exponential growth in situations like the calculation of compound interest, to fail to recognise the phenomenon'
Foltice, B., & Langer, T. (2018). Exponential growth bias matters: Evidence and implications for financial decision making of college students in the U.S.A. *Journal of Behavioral and Experimental Finance, 19*, 56–63. https://doi.org/10.1016/j.jbef.2018.04.002

105 Page 270 'economists Matthew Levy and Joshua Tasoff presented subjects with questions'
Levy, M., & Tasoff, J. (2016). Exponential-Growth Bias and Lifecycle Consumption. *Journal of the European Economic Association, 14*(3), 545–83. https://doi.org/10.1111/jeea.12149

106 Page 271 'the people who performed worst on the test were also those who were most confident in their answers'
Ibid.

107 Page 271 'leaving themselves without proper provision for their old age'
Goda, G. S., Levy, M., Manchester, C. F., Sojourner, A., & Tasoff, J. (2015). *The Role of Time Preferences and Exponential-Growth Bias in Retirement Savings.* https://doi.org/10.3386/w21482

108 Page 272 'double an individual's debt-to-income ratio'
Stango, V., & Zinman, J. (2009). Exponential Growth Bias and Household Finance. *Journal of Finance, 64*(6), 2807–49. https://doi.org/10.1111/j.1540-6261.2009.01518.x

109 Page 272 'a significant impediment to the implementation of effective strategies to control infectious disease'
Lammers, J., Crusius, J., & Gast, A. (2020). Correcting misperceptions of exponential coronavirus growth increases support for social distancing. *Proceedings*

of the National Academy of Sciences of the United States of America, 117(28), 16264–6. https://doi.org/10.1073/pnas.2006048117

110   Page 272 'unable to see the importance of disease-control mitigations and hence less likely to implement or observe them'
Ibid.

111   Page 273 'more prone to underestimating the absolute growth rate of the epidemic than liberals'
Ibid.

112   Page 273 'more likely to comply with suggested protective behaviour'
Ibid.

113   Page 278 'likely that we will still see global temperatures rise by at least 1.5 degrees'
Zhou, C., Zelinka, M. D., Dessler, A. E., & Wang, M. (2021). Greater committed warming after accounting for the pattern effect. *Nature Climate Change*, 11(2), 132–6. https://doi.org/10.1038/s41558-020-00955-x

114   Page 280 'The Snowball Earth hypothesis'
Kirschvink, J. L. (1992). Late Proterozoic low-latitude global glaciation: the snowball Earth. *The Proterozoic Biosphere*, 52.

115   Page 285 'limited number of studies which purport to provide evidence that nominative determinism is a real phenomenon'
Keaney, J. J., Groarke, J. D., Galvin, Z., McGorrian, C., McCann, H. A., Sugrue, D., Keelan, E., Galvin, J., Blake, G., Mahon, N. G., & O'Neill, J. (2013). The Brady Bunch? New evidence for nominative determinism in patients' health: Retrospective, population based cohort study. *BMJ (Online)*, 347. https://doi.org/10.1136/bmj.f6627

116   Page 285 'Perhaps the most amusing of these'
Limb, C., Limb, R., Limb, C., & Limb, D. (2015). Nominative determinism in hospital medicine. *The Bulletin of the Royal College of Surgeons of England*, 97(1), 24–6. https://doi.org/10.1308/147363515X14134529299420

117   Page 286 'the highest proportion of names'
Ibid.

118   Page 286 'in comparison to other names in the same profession'
Pelham, B. W., Mirenberg, M. C., & Jones, J. T. (2002). Why Susie sells seashells by the seashore: Implicit egotism and major life decisions. *Journal of Personality and Social Psychology*, 82(4). https://doi.org/10.1037/0022-3514.82.4.469

119 Page 286 'as the underlying causative reason'
Ibid.

120 Page 287 'Critics argued that the comparison wasn't a fair one'
Simonsohn, U. (2011). Spurious? Name similarity effects (implicit egotism) in marriage, job, and moving decisions. *Journal of Personality and Social Psychology, 101*(1). https://doi.org/10.1037/a0021990

121 Page 287 'you would be more likely to find men named Dennis'
Ibid.

122 Page 287 'the Maryland researchers went back to the drawing board'
Pelham, B., & Mauricio, C. (2015). When Tex and Tess Carpenter Build Houses in Texas: Moderators of Implicit Egotism. *Self and Identity, 14*(6), 692–723. https://doi.org/10.1080/15298868.2015.1070745

123 Page 287 'They found similar evidence for nominative determinism'
Ibid.

124 Page 289 'Karl Popper referred to self-fulfilling prophecies as the 'Oedipus effect' in his early and influential work'
Popper, K. (2013). *The Poverty of Historicism.* Routledge. https://doi.org/10.4324/9780203538012

125 Page 290 'increase the likelihood of rejection by colleagues'
Marr, J. C., Thau, S., Aquino, K., & Barclay, L. J. (2012). Do I want to know? How the motivation to acquire relationship-threatening information in groups contributes to paranoid thought, suspicion behavior, and social rejection. *Organizational Behavior and Human Decision Processes, 117*(2), 285–97. https://doi.org/10.1016/j.obhdp.2011.11.003

126 Page 291 'found that Americans of Chinese and Japanese heritage were more likely to die of heart problems on the fourth day of each month than on any other day'
Phillips, D. P., Liu, G. C., Kwok, K., Jarvinen, J. R., Zhang, W., & Abramson, I. S. (2001). The *Hound of the Baskervilles* effect: Natural experiment on the influence of psychological stress on timing of death. *BMJ, 323*(7327), 1443–6. https://doi.org/10.1136/bmj.323.7327.1443

127 Page 296 'the same phenomenon may be at play with sniffer dogs'
Lit, L., Schweitzer, J. B., & Oberbauer, A. M. (2011). Handler beliefs affect scent detection dog outcomes. *Animal Cognition, 14*(3), 387–94. https://doi.org/10.1007/s10071-010-0373-2

128   Page 297 'whether laboratory animals could be bred to be smarter'
Rosenthal, R., & Fode, K. L. (1963). The effect of experimenter bias on the performance of the albino rat. *Behavioral Science*, 8(3), 183–9. https://doi.org/10.1002/bs.3830080302

129   Page 297 'an other-imposed self-fulfilling prophecy'
Ibid.

130   Page 298 'carry out another pioneering experiment, this time on Jacobson's schoolchildren and their teachers'
Rosenthal, R., & Jacobson, L. (1968). Pygmalion in the classroom. *The Urban Review*, 3(1), 16–20. https://doi.org/10.1007/BF02322211

131   Page 298 'an other-imposed self-fulfilling prophecy'
Ibid.

132   Page 300 'no evidence that wind turbines cause any of these conditions has been found'
Knopper, L. D., & Ollson, C. A. (2011). Health effects and wind turbines: A review of the literature. *Environmental Health: A Global Access Science Source* (10)78. https://doi.org/10.1186/1476-069X-10-78

133   Page 300 'those that have been exposed to the most negative information about the detrimental impact of wind farms'
Chapman, S., St. George, A., Waller, K., & Cakic, V. (2013). The Pattern of Complaints about Australian Wind Farms Does Not Match the Establishment and Distribution of Turbines: Support for the Psychogenic, 'Communicated Disease' Hypothesis. *PLoS ONE*, 8(10). https://doi.org/10.1371/journal.pone.0076584

134   Page 300 'the framing of the discourse surrounding the noise'
Crichton, F., Dodd, G., Schmid, G., Gamble, G., & Petrie, K. J. (2014). Can expectations produce symptoms from infrasound associated with wind turbines? *Health Psychology*, 33(4), 360–4. https://doi.org/10.1037/a0031760

135   Page 300 'compared to pre-exposure levels'
Ibid.

136   Page 307 'a significant proportion of their household's income'
Kerchoff, A. C. (1982). Analyzing a Case of Mass Psychogenic Illness. In M. J. Colligan, J. W. Pennebaker, & L. R. Murphy (Eds.), *Mass Psychogenic Illness* (First, pp. 5–21). Routledge.

Chapter 8

137 Page 313 'it did little to dissuade others'
Pan, J., & Siegel, A. A. (2020). How Saudi Crackdowns Fail to Silence Online Dissent. *American Political Science Review, 114*(1), 109–25. https://doi.org/10.1017/S0003055419000650

138 Page 315 'As part of a pair of studies into reactance, students were subjected to health messages'
Dillard, J. P., & Shen, L. (2005). On the Nature of Reactance and its Role in Persuasive Health Communication. *Communication Monographs, 72*(2), 144–68. https://doi.org/10.1080/03637750500111815

139 Page 315 'Another study looked into the impact of newly placed "No Diving" signs positioned at the shallow end of school swimming pools'
Goldhber, G. M., & deTurck, M. A. (1989). A Developmental Analysis of Warning Signs: The Case of Familiarity and Gender. *Proceedings of the Human Factors Society Annual Meeting, 33*(15), 1019–23. https://doi.org/10.1177/154193128903301525

140 Page 316 'researchers found the messaging actually increased the subjects' desire to smoke'
Hyland, M., & Birrell, J. (1979). Government Health Warnings and the 'Boomerang' Effect. *Psychological Reports, 44*(2), 643–7. https://doi.org/10.2466/pro.1979.44.2.643

141 Page 316 'led to a significant increase in smoking in comparison to teenagers not exposed to the messaging'
Robinson, T. N., & Killen, J. D. (1997). Do Cigarette Warning Labels Reduce Smoking? Paradoxical Effects Among Adolescents. *Archives of Pediatrics and Adolescent Medicine, 151*(3), 267–72. https://doi.org/10.1001/archpedi.1997.02170400053010

142 Page 316 'might outweigh the meagre benefits in consumer knowledge gained through such messaging'
Ringold, D. J. (2002). Boomerang Effects in Response to Public Health Interventions: Some Unintended Consequences in the Alcoholic Beverage Market. *Journal of Consumer Policy*. Kluwer Academic Publishers. https://doi.org/10.1023/A:1014588126336

143   Page 325 'algorithms able to beat Grand Masters in the ancient Chinese game
      of Go'
      Silver, D., Huang, A., Maddison, C. J., Guez, A., Sifre, L., Van Den Driessche, G.,
      . . . Hassabis, D. (2016). Mastering the game of Go with deep neural networks
      and tree search. *Nature, 529*(7587), 484–9. https://doi.org/10.1038/nature16961

144   Page 325 'take professional poker players to the cleaners'
      Brown, N., & Sandholm, T. (2019). Superhuman AI for multiplayer poker.
      Science. American Association for the Advancement of Science. https://doi.
      org/10.1126/science.aay2400

145   Page 326 'an accuracy that rivals that of human radiologists'
      McKinney, S. M., Sieniek, M., Godbole, V., Godwin, J., Antropova, N., Ashrafian,
      H., . . . Shetty, S. (2020). International evaluation of an AI system for breast
      cancer screening. *Nature, 577*(7788), 89–94. https://doi.org/10.1038/s41586-
      019-1799-6

146   Page 326 'rather than finding the more difficult-to-identify animals themselves'
      Geirhos, R., Jacobsen, J. H., Michaelis, C., Zemel, R., Brendel, W., Bethge,
      M., & Wichmann, F. A. (2020). Shortcut learning in deep neural networks.
      Nature Machine Intelligence, 2(11), 665–73. https://doi.org/10.1038/s42256-
      020-00257-z

147   Page 327 'The simple addition of a few small stickers has tricked computer-
      vision algorithms for self-driving cars into misreading signs that should have
      told the driver to "Stop" at speed-limit signs'
      Eykholt, K., Evtimov, I., Fernandes, E., Li, B., Rahmati, A., Xiao, C., . . . Song, D.
      (2018). Robust Physical-World Attacks on Deep Learning Visual Classification.
      In *Proceedings of the IEEE Computer Society Conference on Computer Vision and
      Pattern Recognition* (pp. 1625–34). https://doi.org/10.1109/CVPR.2018.00175

148   Page 328 'One deep learning algorithm, designed to diagnose pneumonia, fell
      into precisely that trap'
      Zech, J. R., Badgeley, M. A., Liu, M., Costa, A. B., Titano, J. J., & Oermann, E. K.
      (2018). Variable generalization performance of a deep learning model to detect
      pneumonia in chest radiographs: A cross-sectional study. *PLOS Medicine,
      15*(11), e1002683. https://doi.org/10.1371/journal.pmed.1002683

149   Page 329 'the hospitals' pneumonia prevalence to make its classification'
      Ibid.

150   Page 334 'musing on the potential food shortages that might result from a rapidly growing population'
Ehrlich, P. R. (1968). *The Population Bomb*. New York: Sierra Club/Ballantine Books.

151   Page 337 'Even under the government's prevailing *mitigation* strategy, the modellers suggested that 250,000 would die from the disease'
Ferguson, N. M., Laydon, D., Nedjati-Gilani, G., Imai, N., Ainslie, K., Baguelin, M., . . . Gaythorpe, K. (2020). Report 9: Impact of non-pharmaceutical interventions (NPIs) to reduce COVID-19 mortality and healthcare demand. *Imperial College COVID-19 Response Team*, (March), 1–20. https://doi.org/https://doi.org/10.25561/77482

152   Page 340 'they were quick to warn of the consequences for global agriculture and human health'
Halocarbons: Effects on Stratospheric Ozone. (1976). National Academy of Sciences.

153   Page 344 'people who perceived that they were not expected to be successful were more likely to do well on performance evaluations'
Nurmohamed, S. (2020). The Underdog Effect: When Low Expectations Increase Performance. *Academy of Management Journal*, 63(4), 1106–33. https://doi.org/10.5465/AMJ.2017.0181

154   Page 344 'In another lab-based study'
Ibid.

155   Page 344 'the seemingly authoritative low expectations'
Ibid.

156   Page 353 'To hear former test pilot Tom Morgenfeld describe it'
Westwick, P. J. (2011). *Oral history interview with Thomas Morgenfeld*. Huntington Library, San Marino, California. https://hdl.huntington.org/digital/collection/p15150coll7/id/45064/

157   Page 353 'As Morgenfeld put it'
Ibid.

158   Page 353 'Morgenfeld recalls'
Ibid.

159   Page 354 'As Morgenfeld recalls'
Ibid.

Chapter 9

160 Page 360 'A 2015 study found that "better-to-wait" sex education did not decrease the rates of STIs'
Petrova, D., & Garcia-Retamero, R. (2015). Effective Evidence-Based Programs For Preventing Sexually-Transmitted Infections: A Meta-Analysis. *Current HIV Research, 13*(5), 432–8. https://doi.org/10.2174/1570162x13666150511143943

161 Page 360 'Abstinence-only sex education does not lead to a decrease in teen pregnancy'
Underhill, K., Montgomery, P., & Operario, D. (2007). Sexual abstinence only programmes to prevent HIV infection in high income countries: systematic review. *British Medical Journal, 335*(7613), 248–52. https://doi.org/10.1136/bmj.39245.446586.BE

162 Page 360 'whereas comprehensive sex education leads to a reduction in teen birth rates'
Fox, A. M., Himmelstein, G., Khalid, H., & Howell, E. A. (2019). Funding for Abstinence-Only Education and Adolescent Pregnancy Prevention: Does State Ideology Affect Outcomes? *American Journal of Public Health, 109*(3), 497–504. https://doi.org/10.2105/AJPH.2018.304896

163 Page 363 'It has been suggested that up to 70 per cent of people may display some degree of normalcy bias during a disaster'
Gutierrez, C. M., O'Neill, M., & Jeffrey, W. (2005). Final Report on the Collapse of the World Trade Center Towers. In *Federal Building and Fire Safety Investigation of the World Trade Center Disaster*. http://www.nist.gov/customcf/get_pdf.cfm?pub_id=861610

164 Page 364 'Some people find themselves in an incongruously calm state of *negative panic* or *behavioural inaction*'
Thompson, J. (2003). Surviving a disaster. *The Lancet, 362*, s56–s57. https://doi.org/10.1016/S0140-6736(03)15079-9

165 Page 364 'only 42.5 per cent of the state's residents in mandatory evacuation areas had actually left their homes'
Kulkarni, P. A., Gu, H., Tsai, S., Passannante, M., Kim, S., Thomas, P. A., Tan, C. G., & Davidow, A. L. (2017). Evacuations as a Result of Hurricane

Sandy: Analysis of the 2014 New Jersey Behavioral Risk Factor Survey. *Disaster Medicine and Public Health Preparedness, 11*(6), 720–8. https://doi.org/10.1017/dmp.2017.21

166 Page 364 'fewer than half of residents in the mandatory evacuation area "Zone A" had left'
Brown, S., Parton, H., Driver, C., & Norman, C. (2016). Evacuation during Hurricane Sandy: Data from a rapid community assessment. *PLoS Currents, 8* (DISASTERS). https://doi.org/10.1371/currents.dis.692664b92a-f52a3b506483b8550d6368

167 Page 364 'A total of 159 people across the East Coast of the US lost their lives as a result of Hurricane Sandy: 43 in New Jersey and 71 in New York state'
Diakakis, M., Deligiannakis, G., Katsetsiadou, K., & Lekkas, E. (2015). Hurricane Sandy mortality in the Caribbean and continental North America. *Disaster Prevention and Management: An International Journal, 24*(1), 132–148. https://doi.org/10.1108/DPM-05-2014-0082

168 Page 364 'The leading cause of death was drowning'
Centers for Disease Control and Prevention (CDC). (2013). Deaths associated with Hurricane Sandy – October–November 2012. *MMWR. Morbidity and Mortality Weekly Report, 62*(20), 393–7. http://www.ncbi.nlm.nih.gov/pubmed/23698603

169 Page 365 'under a mandatory evacuation order'
Ibid.

170 Page 365 'associated with higher evacuation rates'
Brown, S., Parton, H., Driver, C., & Norman, C. (2016). *PLoS Currents, 8* (DISASTERS). https://doi.org/10.1371/currents.dis.692664b92af52a3b506483b8550d6368

171 Page 365 'Hurricane Sandy caused $65 billion worth of damage across the US, making it the second-most costly Atlantic hurricane in US history at the time'
Costliest U.S. tropical cyclones tables updated. (2018). https://www.nhc.noaa.gov/news/UpdatedCostliest.pdf

172 Page 365 'one survey of residents of New Jersey found that only 54% said they would actively prepare themselves in advance for another hypothetical storm in the future'
Burger, J., Gochfeld, M., & Lacy, C. (2019). Concerns and future preparedness

plans of a vulnerable population in New Jersey following Hurricane Sandy. *Disasters, 43*(3), 658–85. https://doi.org/10.1111/disa.12350

173   Page 365 'suggested they would simply "be prepared"'
Ibid.

174   Page 368 'One study went so far as to suggest that cows tend to stand up in warm weather, to expose a larger surface area, so as to cool themselves effectively'
Allen, J. D., & Anderson, S. D. (2013). Managing Heat Stress and its Impact on Cow Behavior. *28th Annual Western Dairy Management Conference*, 150–62.

175   Page 372 'The magnitude-6.9 Loma Prieta earthquake which struck central California in 1989 caused almost a billion dollars' worth of insured damage'
Grossi, P., & Zoback, M. L. (2009). *Catastrophe Modeling and California Earthquake risk: a 20-year perspective. Special report.* https://forms2.rms.com/rs/729-DJX-565/images/eq_loma_prieta_20_years.pdf

176   Page 372 'Instead, insurance stock prices increased in the immediate aftermath of the earthquake'
Shelor, R. M., Anderson, D. C., & Cross, M. L. (1992). Gaining from Loss: Property-Liability Insurer Stock Values in the Aftermath of the 1989 California Earthquake. *The Journal of Risk and Insurance, 59*(3), 476. https://doi.org/10.2307/253059

177   Page 372 'People's heightened perception of risk'
Rodrigue, C. M. (1995). Earthquake Insurance: A Longitudinal Study of California Homeowners by Risa Palm. *Yearbook of the Association of Pacific Coast Geographers, 57*(1), 191–5. https://doi.org/10.1353/pcg.1995.0008

178   Page 372 'more than enough to cover the payouts'
Shelor, R. M., Anderson, D. C., & Cross, M. L. (1992). *The Journal of Risk and Insurance, 59*(3), 476.

179   Page 373 'In 2006, the number of active flood- insurance policies increased by over 14 per cent'
Michel-Kerjan, E., Lemoyne de Forges, S., & Kunreuther, H. (2012). Policy Tenure Under the U.S. National Flood Insurance Program (NFIP). *Risk Analysis, 32*(4), 644–58. https://doi.org/10.1111/j.1539-6924.2011.01671.x

180   Page 375 'the predictions of commercial forecasters were biased towards an increased chance of rain'
Bickel, J. E., & Kim, S. D. (2008). Verification of The Weather Channel probability

of precipitation forecasts. *Monthly Weather Review, 136*(12), 4867–81. https://doi.org/10.1175/2008MWR2547.1

181 Page 376 'the ambiguity associated with such forecasts'
Ibid.

182 Page 380 'for which we can forecast with a given degree of accuracy improves by a day'
Bauer, P., Thorpe, A., & Brunet, G. (2015). The quiet revolution of numerical weather prediction. *Nature 525*(7567), pp. 47–55. https://doi.org/10.1038/nature14956

183 Page 380 'a few million atoms'
Freddolino, P. L., Arkhipov, A. S., Larson, S. B., McPherson, A., & Schulten, K. (2006). Molecular Dynamics Simulations of the Complete Satellite Tobacco Mosaic Virus. *Structure, 14*(3), 437–49. https://doi.org/10.1016/j.str.2005.11.014

184 Page 380 'a few hundred microseconds'
Lindorff-Larsen, K., Piana, S., Dror, R. O., & Shaw, D. E. (2011). How Fast-Folding Proteins Fold. *Science, 334*(6055), 517–20. https://doi.org/10.1126/science.1208351

185 Page 381 'a thought experiment that subsequently became known as *Laplace's demon*'
Laplace, P. S. (1814). *Essay Philosophique sur les Proabilités.* Gauthier-Villars.

186 Page 383 Sensitive dependence on initial conditions does not completely capture the mathematically important features of chaos. There are systems which exhibit sensitive dependence on initial conditions, but which are not chaotic. A more complete characterization of chaotic behaviour would include a concept known as topological mixing. For our purposes, however, we will stick with this sensitive dependence on initial conditions which chaotic systems should exhibit over sufficiently long time periods.

187 Page 385 'You can hear chaos for yourself'
Cahalan, R. F., Leidecker, H., & Cahalan, G. D. (1990). *Computers in Physics,* 4(4), 368.

188 Page 385 'the variations in animal population sizes'
Rogers, T., Johnson, B., & Munch, S. (2022). *Chaos is not rare in natural ecosystems. Nature Ecology & Evolution,* 6(8):1105-1111. https://doi.org/10.1038/s41559-022-01787-y

189   Page 386 'this horizon is between one and two weeks'
      Hoskins, B. (2013). The potential for skill across the range of the seamless
      weather-climate prediction problem: a stimulus for our science. *Quarterly
      Journal of the Royal Meteorological Society*, 139(672), 573–84. https://doi.
      org/10.1002/qj.1991

190   Page 389 'to determine it must be there'
      Smart, W. M. (1946). John Couch Adams and the discovery of Neptune. *Nature*,
      158(4019), 648–52.

191   Page 391 'what we now consider to be the mathematical subject of chaos theory'
      Lorenz, E. N. (1963). Deterministic Nonperiodic Flow. *Journal of the Atmospheric
      Sciences*, 20(2), 130–41. https://doi.org/10.1175/1520-0469(1963)020<0130:dnf
      >2.0.co;2

192   Page 392 'three variables to understand it better'
      Ibid.

193   Page 392 'He shared his discoveries in a paper'
      Lorenz, E. N. (1972). Does the flap of a butterfly's wings in Brazil set off a
      tornado in Texas? *American Association for the Advancement of Science.*